Laboratory Manual of Biochemistry
Methods and Techniques

www.nipaers.com
Browse, Search, Read & Buy...
Print Books eChapters
eBooks Publishers
Forthcoming Subject Catalogues
New Titles
Online Resources on:
✓ Current Affairs
✓ Reasoning, Logic & Aptitude
✓ Competitive Examinations
✓ English Language Lab
✓ Writing & Pronunciation Tools
✓ Personality Development
✓ Online Programmes for Professionals
✓ Interview Preparation
Pay Using
PhonePe Google Pay paytm PayPal UPI CCAvenue

Laboratory Manual of Biochemistry
Methods and Techniques

For B.Sc., B.Sc. Ag, BV.Sc, B.Sc. Biotechnology, B.Tech. Biotech
and
M.Sc., M.Sc. Ag, MV. Sc M.Sc. Biochemistry, M.Sc. Biotechnology and M.Tech. Biotech
Strictly according to new syllabus prescribed by UGC, ICAR and All India Technical Universities

R.S. Sengar
and
Reshu Chaudhary
Department of Biotechnology
College of Agriculture
S.V. Patel University of Agriculture and Technology
Meerut 250 110, U.P.

NEW INDIA PUBLISHING AGENCY
New Delhi – 110 034

NEW INDIA PUBLISHING AGENCY

101, Vikas Surya Plaza, CU Block, LSC Market
Pitam Pura, New Delhi 110 034, India
Phone: + 91 (11)27 34 17 17 Fax: + 91(11) 27 34 16 16
Email: info@nipabooks.com
Web: www.nipabooks.com

Feedback at feedbacks@nipabooks.com

ISBN: 978-93-83305-02-5

Composed and Designed by NIPA

डा. आर.एस.कुरील
कुलपति

Dr. R.S. Kureel
Vice-Chancellor

नरेन्द्र देव कृषि एवं प्रौद्योगिक विश्वविद्यालय,
कुमारगंज, फैजाबाद-224 229
(उ.प्र.),भारत
Narendra Deva University of Agriculture &Technology
Kumarganj, Faizabad-224 229 (U.P.), India

Foreword

Biochemistry is both a basic and applied science. In fact modern biochemistry has two branches: descriptive biochemistry and dynamic biochemistry. Descriptive biochemistry is concerned with the qualitative and quantitative characterization of the various cell components and the dynamic biochemistry deals with the elucidation of the nature and the mechanism of the reactions involving these cell components. While the former branch is more a concern of the organic chemistry, the latter branch has now become the language of modern biochemistry.

Biochemistry has a great impact on other biological sciences, such as biotechnology, agriculture, physiology, microbiology, medicine, pathology, genetics etc. Rapid advances and improvements in our life processes have been made because of understanding of these in terms of chemical phenomena with the development of newer and newer laboratory techniques and methodologies evolved. Today, biochemistry is considered as an experimental science. This book entitled **"Laboratory Manual of Biochemistry: Methods and Techniques"** will serve as a very useful manual and I congratulate Dr. R.S. Sengar and Dr. Reshu Chaudhary for taking immerse pains in bringing out this book at a right time.

This practical book on basic and applied techniques has been comprehensively for imparting knowledge to the students for teaching biochemistry, physiology and environmental sciences. I hope the student will find this manual useful.

The author's need to be congratulated for bringing out publication.

R.S. Kureel

Preface

Biochemistry is the science which deals with the chemical composition, synthesis, degradation and interaction of compounds found in living organisms. More simply it encompasses understanding of life processes in terms of chemical phenomena. The theme of biochemistry is to correlate the molecular structures with biological functions. Today, the growth of all other biological sciences such as biotechnology, biology, physiology, genetics, pathology, microbiology, medicine, agriculture *etc.* is dependent upon biochemistry. Biochemistry has an important role to play in all the life processes and technological developments in the area of food energy public health and pollution control in order to meet the requirement with the rapidly increasing world population in the years to come. Biochemistry is an experimental science. Rapid advances in our knowledge of life processes have been made because of the development of newer laboratory techniques and appropriate methodologies which yield satisfactory results. Although there are quite a number of books available on general and clinical biochemistry, the laboratory techniques and biochemical methodologies are found scattered in various journals and others. Hence, it is laborious and time consuming for the students and scientists so as to search for a desirable method of analysis. Obviously, there is a need for a book which at one place could provide the various biochemical methods of analysis that are in frequent use in research and class works.

This practical laboratory manual entitled "Biochemistry : Methods and Techniques" has been written to help researchers, teachers and students for proper understanding and use of the major techniques and methods employed in biochemistry. The book is primarily intended for the students, scholars and scientists engaged in agricultural research, working in areas like plant biochemistry, plant breeding, plant biotechnology, genetics, plant physiology, plant pathology, agricultural microbiology, soil science, seed science, agricultural chemistry, food sciences, agricultural entomology and other related sciences. However, this manual is designed to meet the requirements of teachers, researchers, and students alike who are engaged in the study of biochemistry.

We feel extremely happy to bring out this practical guide on experimentation in the field of biochemistry. It is quite certain that it would serve the purpose of being useful to students as a dependable laboratory manual. This book has been prepared on course basis inclusive of some advanced techniques that we have standardized. The chemical composition of different media and their preparations are presented in the appendices. There are 88 exercises in this manual. The exercise contains a number of illustrations that complement the text. It is hoped that the manual would not only be educative but would also serve an as ready reckoner for readers intrigued by this subject.

Besides the above 88 exercises some definitions; weights and measures; preparation of buffers; standard solutions; indicators; stains *etc.* and other relevant information have been inpregnated in the appendices.

In the end we owe the responsibility for mistakes, omissions *etc*, if any and would be highly thankful to those bringing out the same to our notice along with the valuable suggestions for further improvement of this laboratory manual.

Authors

Acknowledgement

I am deeply indebted to Prof. D.P. Singh the Vice-chancellor, Prof. Shivendra Vikram Shahi, Western Kentucky University. Bowling Green, USA, Prof. V.P. Singh, Dean, Dept. of Plant Science, M.J.P. Rohilkhand University, Bareilly and Late Prof. H.S. Srivastava, M.J.P. Rohilkhand University, Bareilly for their encouragement and desire to publish this experimental book for the use of students, scholars, teachers and researchers working in agriculture and other related sciences.

I wish to thank my department colleagues, students and friends for their desire and support to complete this book. I extend very special thanks and gratefully acknowledge the information in the form of figures, tables and others which have been adapted in the book from various sources.

I express my gratitude to all who have helped me directly or indirectly in writing this book, especially, my parents who always inspired me with the following words "Try to leave behind your footprints on the sands of time through some of your humble contributions to the society".

I am thankful to Late Prof. H.S. Srivastava and other respected senior teachers who have imparted their knowledge about the subject to me. This would not have been a successful effort without their blissful inspiration and guidance.

I also thank the editorial staff members of New India Publishing Agency, New Delhi for their active cooperation. And finally I express my sincere gratitude to my friends and near and dear ones for their unfailing emotional support during the process of writing this laboratory manual.

Finally, this manual would never have been accomplished without the encouragement and inspiration of my reverend mother Smt. Kamla Sengar and father Dr. S.S. Sengar, Brother Dr. Rajesh Singh Sengar, my wife Sarita Sengar and great pleasure obtained from the similing faces of my two sons Divyanshu Sengar and Kartikey Sengar while preparing this book.

Authors

Contents

Abbreviations/Acronyms

A	Adenine	min	minute(s)
AAT	Aspartate Amino Transferase	ml	milliliter(s)
Ab	Antibody	MOPS	3-(N-Morphino) ethanesulphonic acid
ABA	Abscisic acid	mRNA	Messenger RNA
ADL	Acid Detergent Lignin	MSI	Micron Separations Inc.
ADP	Adenosine 5' diphosphate	MW	molecular weight
Ag	Antigen	Mr	Relative Molecular mass
AMP	Adenosine 5' monophosphate	NAD^+	Nicotinamide adenine dinucleotide (oxidised)
Amp	Ampicillin	$NADP^+$	Nicotinamide adenine dinucleotide phosphate (oxidised)
APS	Ammonium Per Sulfate	NADPH	Nicotinamide adenine dinucleotide phosphate (reduced)
ATP	Adenosine 5' triphosphate	NaOAc	Sodium Acetate
ATPase	Adenosine triphosphatase	NFE	Nitrogen-Free-Extract
BME	β-mercaptoethanol	Ng	Nanogram(s) = 10^{-9} gram
Bp	Base pair	NG	N-methyl-N-nitro-N-nitrosoguanidine
BPB	bromophenol blue	Nif genes	Nitrogen-fixing genes
BSA	bovine serum albumine CSPD, Disodium 3-(4 methoxyspiro {1,2- dioxetane-3, 2' (5'-chloro) tricycle [3.3.1.13,7]decan}-4-yl) phenyl phosphate	NiR	Nitrite reductase
BV	Biological Value	NNED	N-(1-naphthyl) ethylenediamine
C	Cytosine	nm	nanometer(s) = 10^{-9} meter
cAMP	Cyclic AMP	NPR	Net Protein Ratio
CI	Chymotrypsin Inhibitor	NPU	Net Protein Utilization
CIU	Chymotrypsin Inhibitor Units	NR	Nitrate Reductase
cpm	Counts per minute	OD	optical density
CTAB	mixed alkyltrimethyl-ammonium	ODx	optical density at x nm

Contd.

CTP	Cytidine tri phosphate	PAL	L-Phenylalanine ammonia lyase
dATP	deoxyadenosine 5'- triphosphate	PBS	Phosphate buffered saline triphosphate
DCPIP	2,6-Dichlorophenol indophenols	PER	Protein efficiency ratio
dCTP	deoxycytidine 5'-triphosphate	PEP	Phosphoenolpyruvate
ddH$_2$O	double-distilled water	PEP-carboxylase	Phosphoenolpyruvate carboxylase
DEAE	Diethylaminoethyl cellulose	PCR	Polymerase chain reaction
dGTP	deoxyguanosine 5'-triphosphate	PG	Polygalactouronase
dH$_2$O	distilled water	Pi	Inorganic phosphate
Dig	Digoxigenin	pI	Isoelectric point
Dig-dUTP	digoxigenin-11-dUTP	PME	Pectin methyl esterase
DMSO	Dimethyl sulphoxide	PMS	Phenazine methosulfate
DNA	deoxyribose nucleic acid	ppm	Parts per million
DNase	Deoxyribonuclease	POD	Peroxidase
DNP	2,4- Dinitrophenol	ppb	Parts per billion
DNPH	2,4-Dinitrophenyl hydrazine	PPi	Inorganic pyro phosphate
DNS	3,5- Dinitrosalicyclic acid	PPO	Polyphenol oxidase or DOPA oxidase
dNTPs	deoxynucleoside 5'triphosphates	PR-proteins	Pathogenesis-related proteins
DTT	Dithiothreitol	PS	Photosystem
dUTP	deoxyuridine 5'-triphosphate	PVP	Polyvinylpyrolidone
e-	Electron	RBC	Red blood corpuscles
EC	Enzyme Commission	RCF	Relative centrifugal force
EDTA	Ethylenediaminetetraacetate	RFLPs	restriction fragment length polymorphisms
EIA	Enzyme Immuno Assay	R_f	Relative rate of flow
ELISA	Enzyme-linked immunosorbent assay	RNA	Ribonucleic acid
Eq. wt.	Equivalent weight	RNase	Ribonuclease
EtBr	Ethidium Bromide	rRNA	Ribosomal RNA
EtOH	Ethanol	rpm	Revolutions per minute
FAD	Flavin Adenine dinucleotide (oxidised)	RT	Room Temperature
FADH$_2$	Flavin Adenine dinucleotide (reduced)	RuBP	Ribulose-1, 5-bisphosphate

Contd.

FCR	Folin-Ciocalteu reagent	RuBP carboxylase	Ribulose-1, 5-bisphosphate carboxylase
FDNB	1-Fluro-2, 4- dinitrobenzene	RXN	reaction(s)
FID	Flame ionization detector	SDS	Sodium dodecyl sulphate
FMN	Flavin mono nucleotide (oxidised)	SDS -PAGE	Sodium dodecyl sulphate-polyacrylamide gel electrophoresis
$FMNH_2$	Flavin mono nucleotide (reduced)	Sec	Second(s)
G	Guanine	SGB	Sample gel buffer
g	gram(s)	SS DNA	Salmon sperm DNA
GA	Gibberellic acid	SSC	Saline sodium citrate
GC	Gas Chromatography	STE	Sodium Tris-EDTA (also TEN)
GDH	Glutamate Dehydrogenase	T	Thymine
GLC	Gas liquid chromatography	TAE	Tris-acetate EDTA (buffer)
GOGAT	Glutamate Synthase or Glutamine synthase 2.0 × 0 glutarate aminotransferase	TAL	Tyrosine Ammonia Lyase
GOT	Glutamate oxaloacetate transaminase	TBE	Tris-borate EDTA
GS	Glutamine Synthetase	TCA	Trichloroacetic acid
GTP	Guanosine tri phosphate	TD	True digestibility
h	hour(s)	TI	Trypsin Inhibitor
HAT	Hypoxanthine aminopterin thymidine medium	TIU	Trypsin Inhibitor Units
HEPES	4(2-hydroxyethyl)-1 piperazine-ethanesulphonic acid	TEMED	N,N,N',N'-Tetramethylethylene diamine
HPLC	High Performance Liquid Chromatography	TLC	Thin-layer chromatography
HRP (buffer)	Horse radish peroxidase	TNE	Tris Sodium (Na) EDTA
HU	Hemagglutination unit	TPP	Thiamine pyrophosphate
IAA	Indole-3-acetic acid	Tris	Tris(hydroxymethyl) amino- methane
IEF	Isoelectric focusing	t-RNA	Transfer RNA
Ig G	Immunoglobulin G	TTC	Triphenyltetrazolium chloride

Contd.

IR	Infra Red	TTE	Triton Tris-EDTA (buffer)
kb	Kilobases	TTP	Thymidine 5'-triphosphate
KOAc	potassium acetate	U	Unit(s) of enzyme
Kw	Ion product of water	U	Uracil
Lb	Leghemoglobin	UTP	Uridine-5-triphosphate
LDH	Lactate dehydrogenase	UV	Ultraviolet
LMP	low melting point	V	Volts
mA	milli Amperes	XC	Xylene cyanole
MAK	Methylated albumin Kieselguhr	°C	degree Celsius
MBTH	3-Methyl-2-benzothiazolone hydrazone hydrochloride	μg	microgram(s) = 10^{-6} gram
MCA	Minimum concentration for agglutination	μl	microliter(s) = 10^{-6} liter

Glossary

Abscess: A circumscribed collection of pus.

Abscisic Acid: A phytohormone (plant hormone) utilized to control: the size of stomatal pores – i.e., the openings in leaves through which plants exchange oxygen and carbon dioxide (and water inadvertently) with the atmosphere; abscision (e.g., shedding of flowers, fruit, etc.); dormancy.

Absolute Configuration: The configuration of four different substituent groups around anasymmetric carbon atom, in relation to D- and L-glyceraldehyde.

Absorbance (A): A measure of the amount of light absorbed by a substance suspended in a matrix. The matrix may be gaseous, liquid, or solid in nature. Most biologically active compounds (e.g., proteins) absorb light in the ultraviolet (UV) or visible light portion of the spectrum. Absorbance is used to quantitate (measure) the concentration of the substance in question (e.g., substance dissolved in a liquid).

Absorption: From the Latin *ab*, away, and *sorbere*, to suck into. The taking-up of nutrients, water, etc. by assimilation (e.g., transport of the products of digestion from the intestinal tract across the cell membranes that comprise the gut, and into the blood).

Abzymes: Catalytic antibodies that are synthetic constructs. They either stabilize the transition state of a chemical reaction or bind to a specific substrate, thereby increasing the reaction rate of that chemical reaction.

Ac-CoA: Abbreviation for Acetyl-coenzyme A. Ac-CoA is a chemical synthesized in cell mitochondria by combining the thiol (molecular group) of coenzyme A with an acetyl group (i.e., from breakdown/digestion of fats, carbohydrates, or proteins).

Acid: A substance that contains hydrogen atom(s) in its molecular structure, with a pH in the range from 0–6, which will react with a base to form a salt. Acids normally taste sour and feel slippery. For example, food product manufacturers often add citric acid, malic acid, fumaric acid, and itaconic acid in order to impart a sharp taste to food products.

Activation Energy: The amount of energy (calories) required to bring all the molecules in one mole of a reacting substance to the transition state. More simply, it may also be viewed as the energy required to bring reacting molecules to a certain energy state from which point the reaction proceeds spontaneously.

Activator: A small molecule that stimulates (increases) an enzyme's catalytic activity when it binds to an allosteric site.

Active Site: The region of an enzyme surface that binds the substrate molecule and transforms the substrate molecule into the new (chemical) product (entity). This site is usually located not on a protruding portion of the enzyme, but rather in a cleft or depression. This establishes a controlled environment in which the chemical reaction may occur.

Active Transport: Cell-mediated, energy requiring translocation of a molecule across a membrane in the direction of increasing concentration (i.e., opposite of natural tendency).

Adenine: A purine base, 6-aminopurine, occurring in ribonucleic acid (RNA) as well as in deoxyribonucleic acid (DNA) and a component of adenosine diphosphate (ADP) and adenosine triphosphate (ATP). Adenine pairs with thymine in DNA and uracil in RNA.

Adenosine Diphosphate (ADP): A ribonucleoside 52 -diphosphate serving as phosphate group acceptor in the cell energy cycle.

Adenosine Monophosphate (AMP): A ribonucleoside 52 -monophosphate that is formed by hydrolysis of ATP or ADP.

Adenosine Triphosphate (ATP): The major carrier of chemical energy in the cells of all living things on this planet. A ribonucleoside 52 -triphosphate functioning as a phosphate group donor in the energy cycle of the cell, ATP contains three phosphate/oxygen molecules linked together. When a phosphatephosphate bond in ATP is broken (hydrolyzed), the energy produced can be used by the cell to carry out its functions. Thus, ATP serves as the universal medium of biological energy storage and exchange in living cells.

Adenylate Cyclase: The enzyme (within cells) that catalyzes the synthesis (manufacture) of cyclic AMP.

Agar: A complex mixture of polysaccharides obtained from marine red algae. It is also called agar-agar. Agar is used as an emulsion stabilizer in foods, as a sizing agent in fabrics, and as a solid substrate for the laboratory culture of microorganisms. Agar melts at 100°C (212°F), and when cooled below 44°C (123°F) forms a stiff and transparent gel. Microorganisms are seeded onto and grown (in the laboratory) on the surface of the gel.

Agarose: A highly purified form of agar used as a stationary phase (substrate) in some chromatography and electrophoretic methods.

Aldose: A simple sugar in which the carbonyl carbon atom is at one end of the carbon chain. A class of monosaccharide sugars; the molecule contains an aldehyde group.

Algae: A heterogeneous (widely varying) group of photosynthetic plants, ranging from microscopic single-cell forms to multicellular, very large forms such as seaweed. All of them contain chlorophyll and hence most are green, but some may be different colors due to the presence of other, overshadowing pigments.

Amino Acid: There are 20 common amino acids, each specified by a different arrangement of three adjacent DNA nucleotides. These are the building blocks of proteins. Joined together in a strictly ordered chain, the sequence of amino acids determines the character of each protein (chain) molecule. The 20 common amino acids are: alanine, arginine, aspartic acid, glutamic acid, glutamine, glycine, histidine, isoleucine, leucine, phenylalanine, proline, serine, threonine, tryptophan, tyrosine, valine, cysteine, methionine, lysine, and asparagine. Note that virtually all of these amino acids (except glycine) possess an asymmetric carbon atom, and thus are potentially chiral in nature.

Ampholyte: A trade name for a mixture of substances with a range of isoelectric points that have high buffering capacity at their isoelectric points.

Amylase: A term that is used to refer to a category of enzymes that catalyzes the chemical reaction in which amylose (starch) molecules are hydrolytically cleaved (broken) to molecular pieces (e.g., the polysaccharides maltose, maltotriose, a-dextrin, etc.). For example, α-amylase is used to break apart corn starch molecules in the first step of manufacturing fructose (sweetener for soft drinks). Since 1857, amylase has been utilized to remove (amylose) starch from woven fabrics in the textile industries. Modern uses of some amylases include enabling the substitution of barley grain for malt in the beer brewing process.

Amylose: The form of starch that consists of unbranched polymers, containing approximately 4000 glucose units per molecule (polysaccharide). It is present in potatoes at 23–29% content (variation is thought to be caused by different growing conditions).

Anabolism: The phase of intermediary metabolism concerned with the energy-requiring biosynthesis of cell components from smaller precursor molecules.

Analyte : The substance or component in a sample that is being measured.

Angstrom (Å): A unit used for measuring lengths shorter than 100 Å such as those of atoms and molecules (nanometer, nm is used for longer dimensions); 1 Å = 10^{-10} meter or 10^{-8} centimeter or 0.1 nanometer.

Anion: Negatively-charged particle or ion.

Antibody: A highly specific protein produced in the blood by B lymphocytes in response to a foreign molecule (antigen) and released into the bloodstream. Often binds to the foreign molecule or cell extremely tightly, thereby inactivating it or marking it for destruction by phagocytosis or complement - induced lysis. Each antibody protein is composed of 4 subunits, two heavy chains and two light chains.

Anticoagulant: A substances that hinders agglutination or clotting of blood cells.

Anticodon: A specific sequence of three nucleotides in a transfer RNA (tRNA), complementary to a codon (also three nucleotides) for an amino acid in a messenger RNA.

Antigen: Also called an immunogen. Any large molecule or small organism whose entry into the body provokes synthesis of an antibody or immunoglobulin (i.e., an immune system response).

Ascorbic Acid: A water-soluble vitamin and antioxidant.

Assay: A test (specific technique) that measures a response to a test substance or the efficacy (effectiveness) of the test substance.

Assimilation: The formation of self cellular material from small molecules derived from food.

Band: A chromatographic zone that is a region where the separated substance is concentrated.

Bioassay: Determination of the relative strength or bioactivity of a substance (e.g., a drug). A biological system (such as living cells, organs, tissues, or whole animals) is exposed to the substance in question and the effect on the living test system is measured.

Biochemistry: The study of chemical processes that comprise living things (systems); the chemistry of life and living matter. Despite the dramatic differences in the appearances of living things, the basic chemistry of all organisms is strikingly similar. Even tiny one-celled creatures carry out essentially the same chemical reactions that each cell of a complex organism (such as man) carries out.

Biological Activity: The effect (change in metabolic activity upon living cells) caused by specific compounds or agents. For example, the drug aspirin causes the blood to thin, i.e., to clot less easily.

Biological Oxygen Demand (BOD): The oxygen used in meeting the metabolic needs of aerobic organisms in water containing organic compounds. Numerically, it is expressed in terms of the oxygen consumed in water at a temperature of 68°F (20°C) during a 5-day period. The BOD is used as an indication of the degree of water pollution.

Boundary: Edge of a zone; as of macromolecule solution next to the solvent.

bp: Common abbreviation for base pair.

Callose: A complex carbohydrate in sieve tubes of sieve tube members; callose is especially abundant in injured sieve tubes.

Callus: An undifferentiated cluster of plant cells that is a first step in regeneration of plants from tissue culture.

Calorie: The amount of heat (energy) required to raise the temperature of one gram of water from 14.5°C (58°F) to 15.5°C (60°F) at a constant pressure of one standard atmosphere. This unit measure of energy is also frequently utilized to express the amount of energy contained within certain foods or animal feeds.

Capacity factor: The ratio of the elution volume of a substance to the void volume in the column.

Carbohydrates (saccharides): A large class of carbon-hydrogen-oxygen compounds. Monosaccharides are called simple sugars, of which the most abundant is D-glucose. It is both the major fuel for most organisms and constitutes the basic building block of the most abundant polysaccharides, such as starch and cellulose. While starch is a fuel source, cellulose is the primary structural material of plants. Carbohydrates are produced by photosynthesis in plants. Most, but not all, carbohydrates are represented chemically by the formula $C_x(H_2O)n$, where *n* is three or higher. On the basis of their chemical structures, carbohydrates are classified as polyhydroxy aldehydes, polyhydroxy ketones, and their derivatives.

Catabolism: Energy-yielding pathway. The phase of metabolism involved in the energyyielding degradation of nutrient (food) molecules.

Cation: Positively-charged particle or ion.

Cell Culture: The *in vitro* (outside of body, in a test tube) propagation of cells isolated from living organisms.

Cell: From the Latin *cella*, which means small room. The fundamental self-containing unit of life. The living tissue of every multicelled organism is composed of these fundamental living units. Certain organisms may consist of only one cell, such as yeast or protein bacteria, protozoa, some algae, and gametes (the reproductive stages) of higher organisms. Larger organisms are subdivided into organs that are relatively autonomous but cooperate in the functioning of that plant or animal. Unicellular (single-cell) organisms perform all life functions within the one cell. In a higher organism (a multicellular organism), entire populations of cells (i.e., an organ) may be designated a particular specialized task (e.g., the heart to facilitate circulation). The cells of muscle tissue are specialized for movement, and those of bone and connective tissue for structural support. While most cells are too small to be seen with the unaided eye, the egg yolk of birds is a single cell, so the egg yolk of an ostrich is the world's largest cell.

Centrifuge: A machine that is used to separate heavier from lighter molecules and cellular components and structures.

Chromatogram: A series of separated bands or zones detected either visually, as in paper chromatographic or thin layer chromatographic separations or indirectly by a detection system. In the latter case, the detection system usually outputs an electrical signal, which is graphically plotted through time, to display the series of separated bards or zones.

Coenzyme: A non-proteinaceous organic molecule required for the action of certain enzymes. The coenzyme contains as part of its structure one of the vitamins. This is why vitamins are so critically important to living organisms. Sometimes the same coenzyme is required by different enzymes involved in the catalysis of different reactions. By analogy, a coenzyme is like a part of a car, such as a tire, which can be identified in and of itself and which can, furthermore, be removed from the car. The car (enzyme), however, must of necessity have the tire in order to carry out its prescribed function. Coenzymes have been classified into two large groups: fat soluble and water soluble. Examples of a few water-soluble vitamins are: thiamin, biotin, folic acid, vitamin C, and vitamin B_{12}. Examples of fat-soluble vitamins are: vitamins A, D, E, and K.

Cofactor: A nonprotein component required by some enzymes for activity. The cofactor may be a metal ion or an organic molecule called a coenzyme. The term cofactor is a general term. Cofactors are generally heat stable.

Colony: A growth of a group of microorganisms derived from one original organism. After a sufficient growth period, the growth is visible to the eye without magnification.

Colony Hybridization: A technique using *in situ* hybridization to identify bacterial colonies carrying inserted DNA that is homologous with some particular sequence (probe).

Convection: Mass or bulk movement of one part of a solution relative to the rest, usually because of density differences.

Counterion: An ion of opposite charge to the one under consideration. Generally a similar ion.

Culture: Any population of cells (bacteria, algae, protozoa, virus, yeasts, plant cells, mammalian cells, etc.) growing on, or in, a medium that supports their growth. Typically used to refer to a population of the cells of a single species or a single strain. A medium which contains only one specific organism (e.g., *E. coli* bacteria) is known as a pure culture. A culture may be preserved (stored alive) by freezing, drying (in which the cells go dormant), subculturing on an agar medium, or other methods.

Culture Medium: Any nutrient system for the artificial cultivation of bacteria or other cells. The medium usually consists of a complex mixture of organic and inorganic materials. For example, the classic culture (growth) medium used for bacteria consists of nutrients (required by that bacteria) plus agar to solidify or semisolidify the nutrient containing mass.

Dalton: A unit of molecular mass; approximately equal to the weight of a single hydrogen atom (1.66 x 10^{-24}g).

Deamination: The removal of amino groups from molecules (e.g., in an animal's food) via the energy-consuming metabolism of excess amino acids eaten by that animal. For example, when livestock are fed more lysine (amino acid) than their body needs in a given day (animals' bodies can only utilize the essential amino acids in precise amounts/ratios of their daily diet), the excess lysine is metabolized to urea and then excreted.

Decarboxylation: The removal of carbon dioxide from a biochemical, as in the conversion of oxalosuccinic acid (C_6) to α-ketoglutaric acid (C_5) in the third step of the Krebs cycle.

Decomposition: (1) The break-up of a chemical substance into two or more simpler sunstances. (2) The breakdown of organic material by microbes.

Degenerate Code: A code in which one amino acid is specified by more than one codon.

Dehydration reaction: Water-losing. The process in which a hydroxyl (OH) group is removed from one subunit of a polymer and a hydrogen (H) group is removed from the other subunit.

Dehydrogenases: Enzymes that catalyze the removal of pairs of hydrogen atoms from their substrates.

Denaturation: Partial or complete change in conformation of a polypeptide, protein, or nucleic acid caused by heating or by exposure to chemicals and usually resulting in the loss of biologic property; usually a irreversible process.

Denitrification: The breaking down of nitrates to nitrites and ultimately to free gaseous nitrogen by soil organisms, especially denitrifying bacteria.

Deoxyribonucleic acid, DNA: A polynucleotide having a specific sequence of deoxyribonucleotide units covalently joined through 3′, 5′- phosphodiester bonds; two of the polymers wind around each other, like the outside and inside rails of a winding staircase; serves as a carrier of genetic information.

Deoxyribonucleotide: A nucleotide containing 2-deoxy-D-ribose as its pentose component.

Deoxyribose: A 5-carbon sugar with one oxygen atom less than the related sugar ribose; a component of deoxyribonucleic acid.

Depurination: Cleavage of the glycosidic bond between C-1′ of deoxyribose and a purine base in DNA; used in Maxam-Gilbert sequence analysis.

Desiccation: The process by which a substance is dried out and the moisture removed; desiccation is oftden carried out in a desiccators, which contains a substance which will take up water, e.g., calcium chloride.

Detoxification: The process by which poisonous substances are rendered less harmful; for example, the liver converts ammonia into the less toxic compound urea via ornithine cycle, and hydrogen peroxide is split into water and oxygen by the enzyme catalase.

Dextrin: A polysaccharide carbohydrate that may form an intermediate step in the hydrolysis of insoluble starch to soluble glucose which is ready for cell respiration, translocation or further synthesis. Dextrins possess adhesive properties and are used as adhesives on paper products.

Dextrose ($C_6H_{12}O_6$): A monosaccharide sugar of wide occurence in plants; also known as glucose.

Disaccharides: Carbohydrates consisting of two covalently linked monosaccharide units; hence *di* for two.

Diastase: An enzyme mixture, common in seeds such as barley; involved in starch hydrolysis; the mixture contains amylases for conversion of starch to maltose and maltase for conversion of maltose and glucose.

Differential centrifugation: separation of cell organelles or other particles of different size by their rates of sedimentation in a centrifugal field.

Diffusion: The net movement of molecules from a region of higher concentration to a region of lower concentration as a result of random, spontaneous molecular motions; diffusion tends to distribute molecules uniformly throughout a medium.

Digestion: The enzymatic process by which food is changed chemically into materials, which are soluble and diffusible, that the cells can assimilate, store, oxidize, or use as nourishment.

Disaccharide ($C_{12}H_{22}O_{11}$): A carbohydrate consisting of two monosaccharide units that are linked by a covalent bond; for example, sucrose (table sugar) is a disaccharide formed by linking a molecule of glucose to a molecule of fructose.

Disulfide Bond: An important type of covalent bond formed between two sulfur atoms of different cysteines in a protein. Disulfide bonds (linkages, bridges) contribute to holding proteins together and also help provide the internal structure (conformation) of the protein.

2,4-dichlorophenoxyacetic acid (2,4-D): A growth substance widely used as a herbicide.

Ecology: The study of the interrelationships between organisms and their environments.

EDTA: Ethylenediamine tetraacetate. An organic molecule which, due to the chemical groups it contains and their juxtaposition within that molecule, is able to chelate (bind) certain other molecules such as divalent metal cations. EDTA thus inhibits some enzymes requiring such ions for activity.

Effluent: Mobile phase that has exited from the column.

Elastin: The protein found in elastin fibres.

Electrode: Substances in contact with a conductor. The substances are connected to a source of an electric field.

Electron: A minute negatively-chrged particle in an atom. The negative charge of one electron exactly balances the positive charge of one proton.

Electrophoresis: A technique for separating molecules based on the differential movement of charged particles through a matrix when subjected to an electric field. The term is usually applied to large ions of colloidal particles dispersed in water. The most important use of electrophoresis (currently) is in the analysis of proteins, and then a technique known as gel electrophoresis is used. Since the proportion of proteins varies widely in different diseases, electrophoresis can be used for diagnostic purposes. Electrophoresis, through agarose or other gel matrices, is a common way to separate, identify, and purify plasmid DNA, DNA fragments resulting from digestion (of DNA) with restriction endonucleases, and RNA. Electrophoresis is also used to study bacteria and viruses, nucleic acids, and some types of molecules, including amino acids.

Electroporation: A process utilized to introduce a foreign gene into the genome of an organism. In 1995, the U.S. company Dekalb Genetics Corp. received a patent for producing genetically engineered corn via introduction of a foreign gene into corn cells via electroporation. Electroporation, also called electroporesis or electropermeabilization, uses a brief direct-current (dc) electrical pulse to cause formation of "micropores" (tiny holes) in the surface of cells or protoplasts suspended in a solution (water) containing DNA sequences (genes).

After the gene(s) enter the cell via the temporarily created micropores, the electrical pulse ceases, and the micropores close so that the gene(s) cannot depart the cell. The cell then incorporates (some) of the new genetic material (genes) into its genetic complement (genome), and creates whatever product (i.e., a protein) the newly-introduced gene codes for.

ELISA (test for proteins): An enzyme-linked immunosorbent assay (hence the acronym) which can readily measure less than a nanogram (10^{-9} g) of a protein. This assay is more sensitive than simple immunoassay (tests) because one of the two antibodies used to bind and quantitate (measure) the protein's antigen, based on two concurrent epitopes within the protein, is attached to an enzyme. The enzyme can rapidly convert an added colorless substrate into a colored product or a nonfluorescent substrate into an intensely fluorescent product (thus enabling finer quantitation).

Eluate: A compound or mixture that has been separated in and exited forms the column.

Eluent : Mobile phase.

Elution volume: The volume of mobile phase required to elute a solute from a chromatographic column.

Elution: Removal of a solute from a stationary phase by passage of a suitable mobile phase.

Endonucleases: A class of enzymes capable of hydrolyzing (breaking) the interior phosphodiester bonds of DNA or RNA chains. As opposed to cleavage (by exonucleases) at the terminal bonds (ends) of a chain.

Enzyme: An organic, protein-based catalyst that is not itself used up in the reaction. It is naturally produced by living cells to catalyze biochemical reactions. Each enzyme is highly specific with regard to the type of chemical reaction that it catalyzes, and to the substances (called substrates) upon which it acts. This specific catalytic activity and its control by other biochemical constituents are of primary importance in the physiological functions of all organisms. Although all enzymes are proteins, they may, and usually do, contain additional non-protein components called coenzymes that are essential for catalytic activity.

Equilibrium: The state of a system in which no further net change is occurring; the free energy is at a minimum.

Exonuclease: An enzyme that hydrolyzes (cuts) only a terminal phosphodiester bond of a nucleic acid.

Experiment: The test of a hypothesis. A successful experiment is one in which one or more alternative hypotheses are demonstrated to be inconsistent with experimental observation and are thus rejected.

Explant: An actively dividing plant tissue that can be induced to produce callus tissue in tissue culture.

3′ end: The end of a nucleic acid that lacks a nucleotide bound at the 3′ position of the terminal residue.

5′ end: The end of a nucleic acid that lacks a nucleotide bound at the 5′ position of the terminal residue.

Fat soluble vitamin: Any of several vitamins, including vitamins A, D, E and K that are soluble in organic solvents but insoluble in water.

Food: An organic substance (carbohydrate, protein or fat) furnishing energy or building material for protoplasm.

Frictional coefficient: A measure of the resistance a particle offers to movement through a solvent.

Fructose ($C_6H_{12}O_6$): A monosaccharide found in many plants; also called fruit sugar.

Functional Group: The specific atom or group of atoms that confers a particular chemical property on a biomolecule.

Fungicide: Any chemical compound toxic to fungi.

Fungus (plural: fungi): Any of a major group of saprophytic and parasitic plants that lack chlorophyll and flowers, including molds, toadstools, rusts, mildews, smuts, ergot, mushrooms *Aqaricus bisporus*, and yeasts. Under certain conditions (temperature, humidity, etc.), some fungi can produce mycotoxins via their metabolism.

Furanose: A simple sugar containing the 5-membered furan ring.

Gel: A colloid, where the dispersed phase is liquid and the dispersion medium is solid.

Gel Filtration: Also known as exclusion chromatography. An effective technique for separating molecules (such as peptide mixtures) on the basis of size. This is accomplished by passing a solution of the molecules to be separated over a column of Sephadex®, for example, which is a polymerized carbohydrate derivative that contains tiny holes. The holes are of such a size that some of the smaller molecules diffuse into them and are in this way retained (held back) while the larger molecules are not able to get into the holes and pass on by the solid phase (Sephadex ®, in this example). This, simplistically, is how separation is effected.

Gel: A network of interacting fibre; or a polymer that is solid but traps large amounts of solvent in pores or channels inside.

Germ Plasm: The total genetic variability to an organism, represented by the total available pool of germ cells or seed.

Gradient elution: An elution system where the solvent composition is varied during the run.

Growth Factor: A specific substance that must be present in the organism's tissues (when *in vivo*) or growth medium (when *in vitro*) in order for the growth-factor-specific cells to grow/multiply

Growth Hormone (GH): A hormone produced by the anterior pituitary gland. This hormone is a protein (somatotropin) and can be obtained from the bodies of animals, or produced by genetically engineered microorganisms. Its major action in humans (human growth hormone) is a generalized stimulation of skeletal growth. However, human growth hormone (HGH) is also known to affect the growth of other tissues, to be important in fat, protein, and carbohydrate metabolism, and to enhance the effects of various other hormones.

HETP-(Height Equivalent to a Theoretical plate) : The number obtained by dividing the column length by the theoretical plate number.

Hormone: A type of chemical messenger (peptide), occurring in both plants and animals, that acts to inhibit or excite metabolic activities (in that plant or animal) by binding to receptors on specific cells to deliver its "message." A hormone's site of production is distant from the site of biological activity (i.e., where the message is delivered).

HPLC: High-performance liquid chromatography.

Immunoassay: The use of antibodies to identify and quantify (measure) substances by a variety of methods. The binding of antibodies to antigen (substance being measured) is often followed by tracers, such as fluorescence or (radioactive) radioisotopes, to enable measurement of the substance.

***In situ*:** In the natural or original position (e.g., inside the body).

***In vitro*:** In an unnatural position (e.g., outside the body, in the test tube). *In vitro* is Latin for in glass. For example, the testing of a substance, or the experimentation in (using) a "dead" cell-free system.

***In vivo*:** Latin for "in living"; e.g., the testing of a new pharmaceutical substance or experimentation in (using) a living, whole organism An *in vivo* test is one in which an experimental substance is injected into an animal such as a rat in order to ascertain its effect on the organism.

Inhibition: The suppression of the biological function of an enzyme or system by chemical or physical means.

Ion: From the Greek *ion,* something that goes. An ion is an atom or molecule possessing a positive or a negative electrical charge. Ions are produced by the dissociation (coming apart) of a (electrolyte) molecule resulting from an electrolyte dissolving in a solution. One example is the dissociation of common table salt (sodium chloride) in water, which results in positively charged sodium ions (called cations) and negatively charged chloride ions (called anions). Ions play critically important roles in many biological processes such as nerve activity.

Ion-Exchange Chromatography: Separation of ionic compounds (which include nucleic acids and proteins) in a chromatographic column containing a polymeric resin (i.e., the stationary phase) having fixed charge groups. The process works in that the charges of the column (stationary phase) interact with the opposite charges of the material dissolved in the solution that is flowing through the column (mobile phase). The charge interaction between the column material and, i.e., the protein has the effect of slowing down the rate of movement of the protein through the column. The other molecules, meanwhile, which do not interact with the column, flow right on through. This constitutes the separation process.

Isocratic elution: Elution with a solvent mixture of constant composition.

Isoelectric focussing: The ordering and concentration of substances according to their isoelectric points.

Isoelectric point: The pH at which a substance has a zero net charge.

Kb: An abbreviation for 1,000 (kilo) base pairs of deoxyribonucleic acid (DNA).

Kd: An abbreviation for kilodalton.

Kilobase Pairs (Kbp): A unit of DNA equals 1,000 base pairs.

Kilodalton (Kd): A unit of mass equal to 1,000 Daltons.

Medium: A substance used to provide nutrients for cell growth. It may be liquid (e.g., broth) or solid (e.g., agar).

Mobile phase: The mixture of solvents that is percolated through the column.

Mobility: The velocity a particle or ion attains for a given voltage. A relative measure of how quickly an ion moves in an electric field.

Molecular sieving: Separation of molecules on the basis of their effective sizes.

Normal phase: A chromatographic mode in which the mobile phase is less polar than the stationary phase.

Northern Blotting: A research test/methodology used to transfer RNA fragments from an agarose gel (e.g., following gel electrophoresis) to a filter paper without changing the relative positions of the RNA fragments (e.g., re electrophoresis separation grid).

Nucleic Acids: A nucleotide polymer. A large, chain-like molecule containing phosphate groups, sugar groups, and purine and pyrimidine bases; two types are ribonucleic acid (RNA) and deoxyribonucleic acid (DNA). The bases involved are adenine, guanine, cytosine, and thymine (uracil in RNA). Nucleic acids are either the specific (genetic) informational molecule (DNA), or act as agent (RNA) in causing that information to be expressed (e.g., as a protein).

Optimum Temperature: The temperature at which the maximum growth occurs or maximal enzymatic activity occurs, or at which any reaction occurs maximally.

Partition: Process by which a solute is distributed between two immiscible phases.

Peak: A band or zone in a chromatogram showing a maximum of concentration between two minima.

Physiology: The branch of biology dealing with the study of the functioning of living things. The materials of physiology include all life: animals, plants, microorganisms, and viruses.

Picogram (pg): 10–12 gram or 10–14 ounce (avoirdupoir).

Plate: A Chromatographic term that refers to a single partitioning unit of the chromatographic system.

Polyacrylamide Gel: A "sieving" gel, that is used in electrophoresis.

Polyacrylamide Gel Electrophoreis (PAGE): A form of chromatography in which molecules are separated on the basis of size and charge. The stationary phase (the polyacrylamide gel) is a polymerized version of acrylamide monomers. The gel looks and feels like Jello™. On a molecular basis it consists of an intertwined and cross-linked mesh of polyacrylamide strings. As can be imagined, there are tiny "holes" in the gel (as in a plastic mesh bag) and with enough cross-linking the size of thc holes begins to approach the size of the molecules that are to be separated. Since some molecules will be larger and some smaller, some of them will be able to pass through the gel matrix more easily than others. This is part of the basis for separation. It should be noted at this point that if the gel is cross-linked enough, and because of this the holes in that gel are smaller than the molecules to be separated, then the molecules will not be able to penetrate into the gel and no separation can occur. The charge on the molecule also plays a role in the separation. Functionally, the gel serves to hold and separate the molecules. Although details are not presented here, after the gel has been prepared (poured and cross-linked), a small amount of the solution containing the molecules to be separated is placed into wells (grooves to hold the liquid) on the gel and the system is subjected to an electric current. Over the course of minutes to hours, molecules bearing different charge/mass separate.

Polymerase Chain Reaction (PCR): A reaction that uses the enzyme DNA polymerase to catalyze the formation of more DNA strands from an original one by the execution of repeated cycles of DNA synthesis. Functionally, this is accomplished by heating and melting double-stranded (hydrogen bonded) DNA into single-stranded (nonhydrogen bonded) DNA and producing an oligonucleotide primer complementary to each DNA strand. The primers bind to the DNA and mark it in such a way that the addition of DNA polymerase and deoxynucleoside triphosphates cause a new strand of DNA to form which is complementary to the target section of DNA. The process described previously is repeated (trait, product, etc.) again and again to produce millions of copies (amplicons) of the desired strand of DNA. PCR and its registered trademarks are the property of F. Hoffmann-La Roche & Co. AG, Basel, Switzerland.

Resolution: The degree of separation between two components by chromatography.

Retention time: The time that has elapsed from the injection of the sample into the chromatographic system to the recording of the peak maximum of the component in the chromatogram.

Reversed phase: A chromatographic mode in which the mobile phase is more polar than the stationary phase.

R_f: A ratio used in paper chromatography, and thin-layer chromatography that is the distance from the origin to the centre of the separated zone divided by the distance from the origin to the solvent front.

Saponification: Alkaline hydrolysis of triacyl glycerols to yield fatty acid salts. The molecules thus produced are known as surfactants (surface active agents), commonly called soap. The process of soap making.

SDS: Sodium dodecyl sulfate. Also known as sodium lauryl sulfate (SLS). A surfactant commonly used in biochemical and biotechnological applications for the solubilization of membrane components and hard-to-solubilize (dissolve) molecules. For example, it is often utilized at high concentration in water solution (e.g., along with potassium acetate) to dissolve plant DNA samples (e.g., when a scientist wants to sequence that sample of plant DNA). The SDS/PA in water solution helps the scientist to separate out contaminants commonly present in samples from plant tissues (polysaccharides, proteins, etc.) because DNA molecules are much more soluble in SDS/PA solution than are those contaminant molecules. Above a critical concentration (CMC), SDS forms micelles in water which are thought to be responsible for its solubilizing action. SDS is also used in such items as shampoo.

Selectivity: The ratios of the capacity factors for two substances measured under identical chromatographic conditions; sometimes termed separation factor.

Southern Blot Analysis: A test that is performed on biological samples such as plant DNA (e.g., to ascertain if "inserted" DNA is present in particular plant cells). Gel electrophoresis is used to separate the DNA fragments according to size, and then those fragments are transferred to a filter (blot). Radiolabeled DNA probes or RNA probes are added, and the ones which are complementary to each of the (separated, on blot) fragments will hybridize to those respective DNA fragments. The location (on the blot) and "radioactive label" of those hybridized probes can then be utilized to determine the nature of the DNA that was in those plant cells.

Southern blot: A Kind of electroblot, in which the substances separated and transferred are nucleic acids.

Stacking: Ordering or arranging and concentrating macro-molecules according to their effective mobilities.

Stationary phase: The portion of separation system that is immobilized in the column.

Support: The particles on which the stationary phase is held.

Theoretical elate number: A number defining the efficiency of the chromatographic column.

Tissue Culture: The growth and maintenance (by researchers) of cells from higher organisms *in vitro*, i.e., in a sterile test tube or petri dish environment which contains the nutrients necessary for cell growth. One use of tissue culture is to produce disease free offspring from certain (valuable, high quality) crop plants. Another use of tissue culture methods is for "embryo rescue" to enable "wide crosses" between two different species of plants. In that procedure, pollen from one plant species (e.g., a wild plant possessing disease resistance) is induced to fertilize a plant from another species (e.g., a domesticated crop). The resultant fertilized plant embryo, which would not grow on its own, is "rescued" via tissue culture methods. Following maturation, that wide cross (i.e., a hybrid plant from two species that normally would not cross) produces fertile seeds on its own without any need for further intervention by man.

Ultracentrifuge: A high-speed centrifuge that can attain revolving speeds up to 85,000 rpm and centrifugal fields up to 500,000 times gravity. The machine is used to sediment (i.e., cause to settle out) and hence separate macromolecules (large molecules) and macromolecular structures in a mixture/solution. In general, a centrifuge is a machine that whirls test tubes around rapidly, like a merry-go-round, to force the heavier suspended materials (in the solutions in the test tubes) to the bottoms of those test tubes before the lighter material.

Ultrafiltration: A (mixture) separation methodology that uses the ability of synthetic semipermeable membranes (possessing appropriate physical and chemical natures) to discriminate between molecules in the mixture, primarily on the basis of the molecules' size and shape. Invented and developed by Dr. Roy J. Taylor in the 1950s and 1960s, ultrafiltration is typically utilized for the separation of relatively high-molecular-weight solutes (e.g., proteins, gums, polymers, and other complex organic molecules) and colloidally dispersed substances (e.g., minerals, microorganisms, etc.) from their solvents (e.g., water).

Units (U): A measure (quantitation) of biological activity of a substance, as defined by various standardized assays (tests).

Vitamin: The modern term descended from the original phrase "vital amine" (or "vitamine"), which was coined by Casimir Funk in the early 1900s.

Void volume: The interstitial volume of the chromatographic column; that is the volume of mobile phase imbibed in the pores and around the stationary phase in a column.

Western Blot Test: A test performed on biological samples such as blood (after centrifugation to remove red blood cells from the blood) to detect AIDS antibodies individually. Gel electrophoresis is used to separate the AIDS antigen proteins of killed (known) AIDS viruses. Next the protein bands (resulting from the gel electrophoresis) are exposed to the blood being tested and (AIDS) antibodies stick to specific individual antigens (bands) which are then identified (as being present in the tested blood) via dyes.

Western blot: A Kind of electro blot, in which the substances separated and transferred are proteins.

Zeta potential: The potential produced by the effective charge of a macromolecule usually taken at the boundary between what is moving with the macromolecule and the rest of the solution.

Zone: A particular region or space within a large one, generally distinguished by some property, such as its occupancy by the protein.

❑❑❑

Unit -1

General Laboratory Principles

Hypothesis become theories and theories attain rank of laws after withstanding rigorous experimental tests. Feasibility of a process is confirmed in the laboratory. Qualitative and quantitative analyses give complete chemical picture of the substance. It is with these considerations in mind we proceed to learn what is there in a laboratory.

1. Laboratory

A laboratory is a workshop for scientist. Here students learn the techniques of the preparation, identification and estimation of chemical substances. Before starting experiment, a student must know from where to get the apparatus required for the given experiment and the placement of the chemicals to be used. A student must know the proper use of each equipment and the precautions to be observed while working in the laboratory. A laboratory is provided with the following fittings with which the student must become familiar from teacher.

i) Demonstration Table

Before starting with the experiments, the teacher gives instructions and demonstrates the concerned experiments on demonstration table. In biochemistry laboratory, no seats are made available to the students, so students stand around demonstration table and note the instructions from the teacher.

ii) Students Working Table

A number of wooden or concrete tables are provided for working. Each seat is provided with:

(a) **Reagent shelves:** Reagents or chemicals to be used are placed on the reagent shelf. These are the reagents which are commonly used. For example, all dilute and concentrated acids such as H_2SO_4, HCl, HNO_3, etc and bases like NaOH, NH_4OH, etc.

(b) **Sinks and water taps:** A sink and a water tap are fitted between every two reagent shelves. On either side of the sink, usually two taps are fitted for supply of water.

(c) **Gas taps:** These taps are fitted on the seats for supply of petroleum gas to the burners. Sometimes kerosene is used for producing gas in place of petrol.

iii) Side Shelves

Usually, there are two big shelves fitted on the walls of the laboratory. Reagents and chemicals, which are less frequently used, are placed in these shelves. Sometimes solid chemicals are placed in a separate shelf.

iv) Fume Cup Board

There is at least one fume cup-board in the corner of the laboratory. All experiments releasing poisonous gases or vapours are performed in this cup-board.

v) Balance Room

It is a small room attached to each laboratory. Here, a number of balances are kept for weighing the substances.

vi) Exhaust Fans

Two exhaust fans are provided at the two corners of the laboratory for the removal of poisonous gases and vapours from the laboratory.

2. Common Laboratory Apparatus

The apparatus which is commonly used by XI class student is described below:

1. **Beakers:** Beakers of different sizes such as 150 ml, 200 ml made of soft glass or corning glass are used for taking various liquids.
2. **Test Tubes:** Test tubes of different sizes are available. Small test tubes used for salt analysis known as centrifuging tubes and boiling tubes are also available.
3. **Conical flask:** It is used in volumetric analysis for carrying out titration.
4. **Funnel:** It is used for filtration or for pouring solutions.

5. **Measuring Flask:** It is used in quantitative analysis when we have to prepare a solution of a particular volume. There are flasks of 50 ml, 100 ml and 250 ml capacity. There is a mark on the stem of the flask upto which the liquid is taken to complete the volume.
6. **Glass-Rod:** It is used for stirring purposes. It is also used as an aid for transferring the liquid into the funnel.
7. **China Dish:** It is a small vessel made of porcelain. It is used in crystallization, for concentrating a solution.
8. **Wire Gauze:** It is placed above the flame of the burner so that the glass vessel being heated does not touch the flame directly and hence is prevented from breaking.
9. **Tripod Stand:** It is used for supporting a china dish or a beaker so that it can be heated from below.

Other apparatus with which a student must familiarize are test tube holder, test tube brush, crucible tongs, spatula, watch glass, clamp stand, burette, pipette, water bath, sand bath and centrifugal machine.

3. Instructions to Work in Laboratory

To work in the laboratory, a student must follow the following rules:

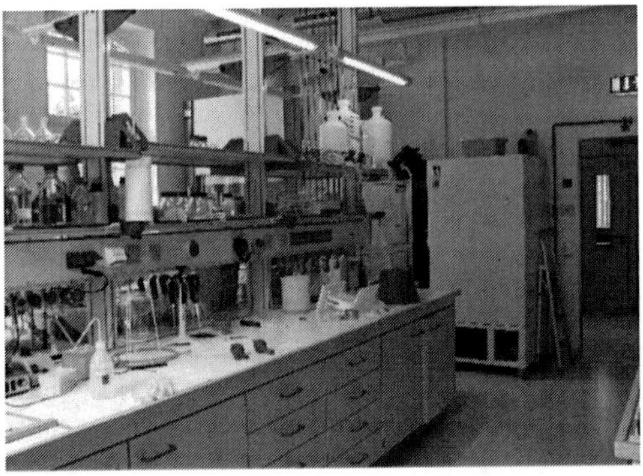

1. A student must have a practical note-book, rough note-book for instructions, a pen or pencil, a laboratory coat and other equipment such as a platinum wire, fractional weights as required.
2. Always come prepared for the experiments. This will help in understanding the experiment better.
3. Always listen to the teacher's instructions carefully and note down the important points and precautions to be followed.
4. After the instructions, collect the apparatus from the laboratory assistant in queue.
5. Thoroughly clean the apparatus to be used.
6. Do only the experiments assigned, unallotted experiments should not be done.
7. Do your experiment honestly without caring for the final result. Record the observations on a rough note-book instead of writing on the pieces of paper.
8. Plan your work so that it is finished in the stipulated time.
9. Be economical with the reagents. Only small quantities of the reagents should be used.
10. Handle the glass apparatus very carefully. In case of any breakage, report it to your teacher at once.
11. Dispose of all waste liquids in the sink allow water to run for sometime by opening the water tap.

12. Keep your seat clean. If an acid or other corrosive chemical is spilled, wash it off with water.
13. Clean your apparatus after the experiments and return it to the laboratory assistant.
14. In case of any injury or accident or breakage of the apparatus, report it to the teacher immediately.
15. Wash your hands with soap after the experiment.

4. Some Important Precautions

To avoid unnecessary risk or injury during laboratory work, the students are advised to observe the following precautions:

1. Do not touch any chemical with hand as some of them may be corrosive.
2. Never taste a chemical. It may be poisonous.
3. Do not place the chemical on the palm of your hand.
4. Do not keep the reagent bottles open.
5. Do not roam here and there in the laboratory without work.
6. Do not put any object into the reagent bottle.
7. Do not bring inflammable liquids such as alcohol, ether near the flame.
8. Do not take the reagent from the shelf to your seat.
9. Do not disturb the arrangement of reagents placed on the shelf.
10. Do not use cracked glass apparatus such as beakers for heating purposes.
11. Do not keep water tap running when not required.
12. Do not throw solid waste materials like filter paper pieces, test-tube pieces, etc. in the sink. Throw them in the waste box only.
13. Do not heat beakers or china dish directly on flame. Always make use of wire gauge.
14. Never add water to the acid. Always pour acid slowly into the water with stirring.
15. Do not play with the chemicals.

5. Practical Note-Book

All the experiments that are conducted in the laboratory are recorded in a practical note-book. It is compilation of whole work done by the student, so it must be well maintained, protected from mechanical and chemical damage. For keeping up-to-date record of experiments following points should be kept in mind.

1. The name of the experiment should be entered along with the date of carrying out that experiment.
2. Requirement should be mentioned next to the title given.
3. Theory and principle of the experiment should be given in a precise manner.
4. This should be followed by the procedure using which experiment is to be conducted. Then a summary of precautions to be taken care are mentioned. Finally mention the general calculations for the experiment.

If we make a table of the points to be written on left hand and right hand side of the note-book, it will look some what like the one given below:

Left hand side	Right hand side
Date	Date
Diagram	Name of experiment
Chemical equation	Theory
Observations	Procedure
Calculations	General calculations Precautions

Keep the following points in consideration regarding your practical note-book:

1. Do not tear pages from note-book.
2. Do not over write if a mistake has been committed in recording, put a line over it and write the correct word or figure again.
3. Number the pages of your note-book.
4. Complete the index, indicating the experiment, its serial number, page number on which it is written.
5. Keep your note-book neat and tidy and covered with brown paper.

The researcher should read and understand the protocol of an experiment before actually starting it. Interpretation of results and trouble-shooting require careful maintenance of experimental records. It is extremely important to always include controls (a negative and a positive control) in an experiment. Every researcher should be aware of the safety considerations of the laboratory. Essential laboratory skills include the ability to prepare solutions and buffers and the practical knowledge of the working principles of the instruments and techniques routinely used in a biology laboratory. This chapter covers most of such general principles of a biology laboratory.

Any laboratory experiment should include maintenance of clear and accurate records of all experiments conducted. It is important to record all experimental findings in a notebook with dates and page numbers. The first few pages can be used for a table of contents. All the details for each experiment can be organized as shown below:

1. **Brief title of experiment and date :**
2. **Aim :**
3. **Procedure :** methodology or a flow chart. Each experiment should ideally have three replicates and appropriate controls (positive and negative controls).
4. **Results :** presented in tabular form or as graphs.
5. **Analysis of data :** the average values are calculated from the replicates and noted.
6. **Inference :** the results obtained should be interpreted in accordance with the principle of the experiment.

6. Laboratory Requirements and Safety

The basic requirements of a biology laboratory include: good computer attached with a printer, autoclaves, sinks, water distillation units, deionizers, hot-air ovens, chemical fume hoods, radiation shields, sterile laminar flow workstations, incubators, temperature control or cold room, dark room for development of photographic films, -20^{0}C freezers, refrigerators, water baths, refrigerated centrifuges, micro centrifuges, pH meter, scintillation counters to monitor radioactivity, trans-illuminator equipped with a camera (to view and photograph gels under UV light) or electronic imaging system, electrophoresis units, blotting apparatus, hot plates, micro wave ovens, balances and many other miscellaneous things.

The laboratory needs a variety of glassware including reagent bottles, beakers, measuring cylinders, Ehrlenmeyer (conical) flasks, test tubes, burettes, pipettes, glass rods, Petri dishes and other culture vessels. Disposable sterile Petri dishes can be used for the culture of microoganisms and plant tissue. Magenta boxes and other autoclavable containers can be used for the culture of plants *in vitro*. Other requirements include liquid nitrogen containers, syringes, needles, forceps, scalpels, membrane filters (to filter sterilize heat labile liquids), magnetic stirrers, orbital shakers, inoculation loops, stop watches, nitro cellulose or nylon membranes, parafilm, saran wrap, aluminium foil, marker pens, Whatman 3 M paper, ice buckets, latex gloves, plastic boxes, plastic bags and U.V goggles, besides plastic bottles and containers. A good supply of chemicals is a very essential aspect of equipping a laboratory as are uninterrupted power and water supply.

Adequate safety measures should be taken to protect the workers like providing safety hoods, good radioactive waste disposal systems, gloves when using hazardous carcinogenic chemicals and wearing of goggles for protection from UV light. Precaution should be taken for containment of pathogenic organisms by autoclaving the culture tubes and glassware before washing them. Other contaminated materials should be rinsed in Chlorox solution for a few minutes before washing them. Working behind a radioactive protection shield when dealing with high-energy radioisotopes such as ^{32}P is mandatory. The work area should be checked for radiation spills by using a Geiger Muller counter.

Exposure to several hazardous and toxic chemicals and other agents in a laboratory poses danger to the researchers so it is essential to adopt safety measures for their protection.

1. Prior to use of an equipment or a chemical the information and instructions should be read.
2. It is essential to the warning signs or labels on equipment & chemicals before using them. It is important to sure that the location of the safety equipment like the eyewashes, first aid kits, cleanup kits and fire extinguishers is known along with the knowledge about their usage.
3. It is mandatory to war lab coats, gloves, eyes protection and inhalation protection masks when working with chemicals, UV light etc.
4. The volatile or potentially hazardous chemicals in a laboratory should be used in a fume hood only.
5. In case of radioactive wastes extreme precaution to dispose them in appropriate containers and clean up spills should be taken immediately. Any radioactive contamination should be checked before and after working in an area of the laboratory, with a Geiger Muller counter. If any radioactivity resists, the problem should be reported to the concerned authorities and efforts should be made to remove it.
6. In case of an injury, medical aid should be sought immediately.

7. First Aid in Laboratory

- In accidental swallowing of acids or alkail, the mouth must be thoroughly rinsed with water. Then in case of acid; rinse the mouth with dilute sodium carbonate of and in case of alkali, with dilute citric acid. If proper swallowing has occurred, the person should be made to drink water followed by milk in case of acids and lemon juice in case of alkali.
- Skin burns should be washed under running water or iced water and then apply petroleum jelly or burn ointment and cover with sterile gauge.
- Inhalation injury by toxic fumes is best treated by shifting the person to an open fresh air atmosphere. Irritation to throat can be soothened with hot water vapour inhalation or a warm drink.

- Chemical injury to eyes must be treated by thorough washing with water and then applying 2% sodium carbonate drops til referred to a specialist.

Additional Reading

Carson P.A. and Dent, N. (eds.) (2007). Good Clinical, Laboratory and Manufacturing Practices: Techniques for the Q A professional. London: RSC. (A comprehensive but easy-to-read book aimed at both newcomers and professional involved in laboratory quantity assurance issues.)

Davenport H.W. (1974). The ABC of Acid-Base Chemistry. 6th ed. University of Chicago Press, Chicago.

God et al. (1966). Biochemistry. 5 : 467.

Robinson J.R. (1961). Fundamentals of Acid-Base Balance. Blackwell, Oxford.

Segel I.H. (1976). Biochemical Calculations. 2nd ed. John Wiley and Sons, Inc., New York.

Safety Control of Substances Hazardous to Health Regulations 2002: Approved Code of Practice and Guidance. Kingston-upon-Thames: HSE Books. (A step-by-step approach to understanding the practical implications of COSHH).

❑❑❑

Unit -2

Carbohydrates

Carbohydrates are synthesized from CO_2 and H_2O in chlorophyll containing plants during photosynthesis. The term carbohydrate was originally coined for this class of compounds as most of them were hydrates of carbon or could be represented by the general formula $C_n(H_2O)_n$. Not all the carbohydrates however, have this empirical formula, for example deoxyribose ($C_5H_{10}O_4$) and glucosamine ($C_6H_{13}O_5N$). Further lactic acid ($C_3H_6O_3$) formaldehyde (HCHO or CH_2O) and acetic acid (CH_{3-}.COOH or $C_2H_4O_2$) with this general formula is not a carbohydrate. Carbohydrates are defined as the compounds having either an aldehyde or a ketone group or a modified aldehyde or ketone group and other carbon atoms with alcoholic (-OH) group. Hence the continued usage of the term carbohydrate is for convenience rather than exactness.

Sugars are small molecules which belong to the class of carbohydrates. As the name implies, a carbohydrate is a molecule whose molecular formula can be expressed in terms of just carbon and water. For example, glucose has the formula $C_6(H_2O)_6$ and sucrose (table sugar) has the formula $C_6(H_2O)_{11}$. More complex carbohydrates such as starch and cellulose are polymers of glucose. Their formulas can be expressed as $C_n(H_2O)_{n-1}$.

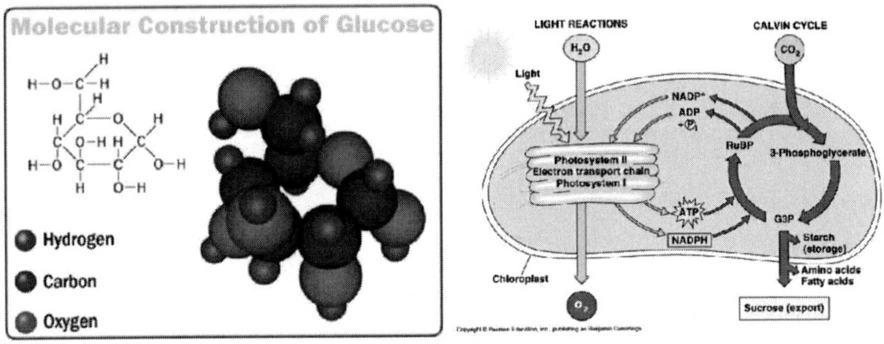

Fig. 2.1 : Molecular Construction of Glucose

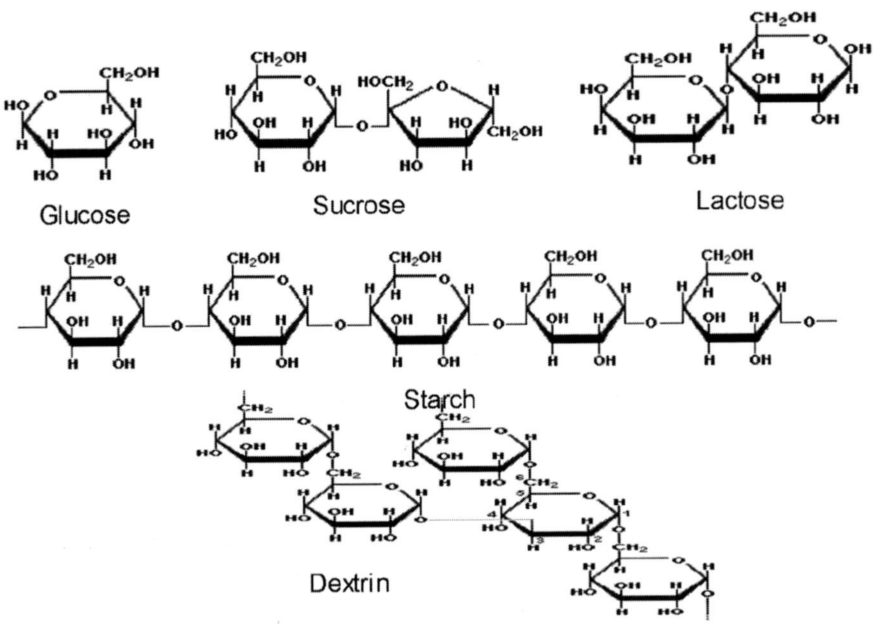

Fig. 2.2 : Carbohydrate Molecules

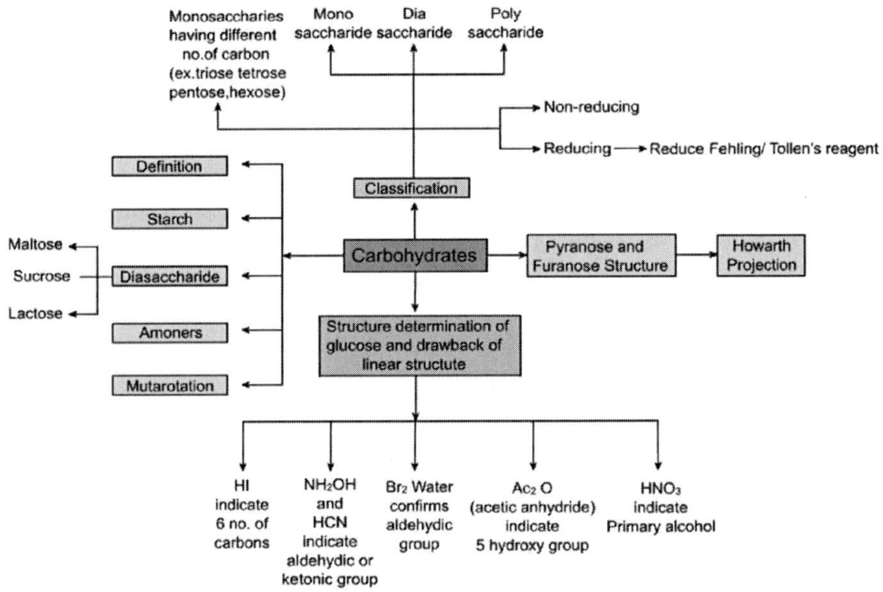

Fig. 2.3 : Classification of Carbohydrate

- *"An experiment is a question which science poses to Nature, and a measurement is the recording of Nature's answer."*

  Max Planck

- *"Knowledge is the edge of desire and can become the pilot of the soul."*

  Will Durant

Experiment – 1 : Analysis of Carbohydrates

Test solutions

1. 0.5 gm glucose in 100 ml SDW
2. 0.5 gm Xylose in 100 ml SDW
3. 0.5 gm Sucrose in 100 ml SDW
4. 0.5 gm Ribose in 100 ml SDW
5. 0.5 gm Lactose in 100 ml SDW

1. Molisch Test

Requirements

Reagents

1. Conc. H_2SO_4 (Solution A)
2. α- napthol - 5% (w/v) in Ethenol (Solution B)

Glass/Plastic wares : Test tube, beaker, glass pipette, test tube stand

Equipments : Balance

Principle

This is a general test for all carbohydrates Conc. H_2SO_4 hydrolysis glycosidic bonds to yield monoaccharides which in the presence of an acid gets dehydrated to form furfural and its derivatives. These products react with sulphonated α-napthol to give a purple complex. Polysaccharides and glycoproteins also give a positive test (reaction).

Reaction

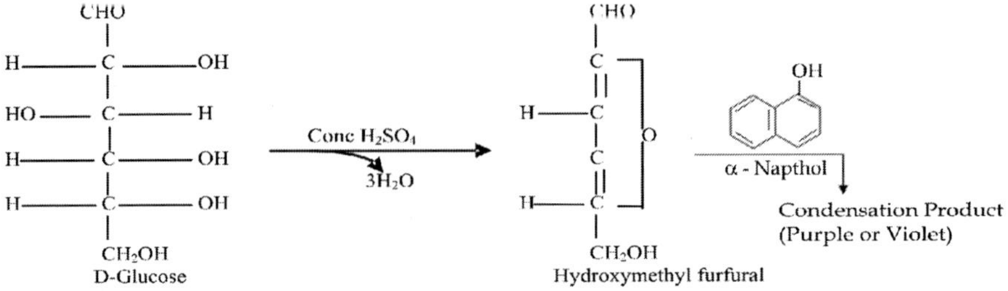

Procedure

1. Take 2 ml test solution & add 2-3 drops of Solution B.
2. Pipette 1 ml Solution A gently along the side of the test tube.
3. Observe the change of colour at the junction of 2 layers.

Observation

Appearance of purple colour indicate positive test for carbohydrates.

2. Anthrone Test

Requirements

Reagents

1. Conc. H_2SO_4 (Solution A)
2. $_2SO_4$ (Solution B)

Glass/Plastic wares : Test tube, beaker, glass pipette, test tube stand

Equipments: Balance

Principle

Anthrone reaction is another general test for Carbohydrates. In this, the furfural produced reacts with anthrone to give bluish green coloured complex.

Reaction

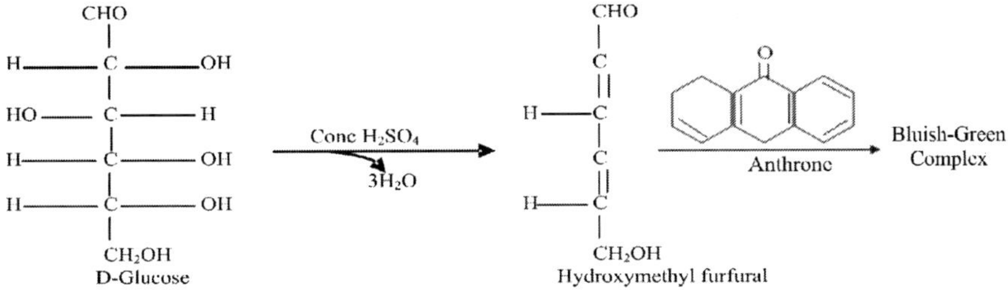

Procedure

1. Take 1 ml of test solution & add 2 ml of solution B.
2. Mix it thoroughly & observe change of colour.

Observation

Appearance of Bluish green indicates positive test for carbohydrates.

3. Iodine Test

Requirements

Reagents : 0.005 NI_2 in 3% KI (Solution A)
Glass/Plastic wares : Test tube, beaker, glass pipette, test tube stand
Equipments: Balance

Principle

Iodine forms coloured adsorption complexes with polysaccharides. Starch gives blue colour with Iodine while glycogen reacts to form reddish brown complex. Hence, it is a useful, convenient & rapid test for detection of amylase, amlyopectin & glycogen.

Procedure

1. Take 1 ml of test solution & add 4-5 drops of Solution A.
2. Mix gently and observe the change of colour.

Observations

Appearance of yellow colour indicates positive test for Carbohydrates.

4. Barfoed's Test

Requirements

Reagents

1. 13.3 g CH_3COOCu in 200 ml SDW (Sol A)
2. 1.00 ml glacial Acetic Acid (Sol B)
3. Mix gently Sol A & Sol B (Barfoed's reagent Sol C)

Glass/Plastic wares : Test tube, beaker, glass pipette, test tube stand
Equipments : Balance

Principle

This test is used for distinguished monosaccharides from reducing disaccharides. Monosaccharides usually react in about 1-2 minute while the reducing disaccharides take much longer time b/w 7-12 min. to get hydrolysed and then react with the reagent. Brick Red Colour is obtained in the test which is due to formation of cuprous oxide.

Reaction

$$\underset{\text{Cupric Acetate}}{(CH_3COO)_2Cu} + 2H_2O \longrightarrow 2CH_3COOH + \underset{\text{Cupric hydroxide}}{Cu(OH)_2}$$

$$Cu(OH)_2 \longrightarrow CuO + H_2O$$

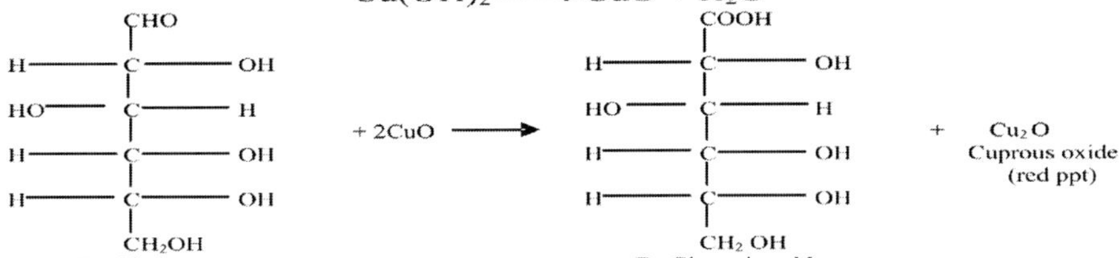

D- Glucose D- Gluconic acid

Procedure

1. Take 1 ml test solution & add 2 ml of Sol C.
2. Cook in Boiling Water Bath.
3. Observe the colour.

Observation

Appearance of Brick red colour indicates positive test for carbohydrates.

5. Seliwanoff's Test

Requirements

Reagents : 0.05% w/v resorcinol in 3N HCl (Sol A)

Glass/Plastic wares : Test tube, beaker, glass pipette, test tube stand

Equipments: Balance

Principle

This test is used to distinguish aldoses from ketoses, ketoses dehydration to give furfural derivatives which then condense with resorcinol to form a red complex. Prolonged heating will hydrolyse disaccharides & other monosaccharides will also eventually give colour.

Reaction

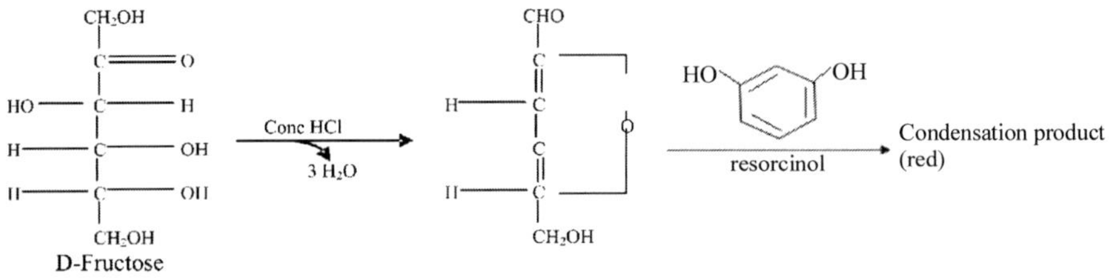

Procedure

1. Take 1 ml test solution & add 2 ml of sol. A.
2. Cook in water Bath for 1 min.
3. Observe the colour.

Observation

Appearance of deep red colour indicates presence of ketose sugar.

6. Fehling's Test

Requirements

Reagents

1. 35.0 g $\cdot CuSO_4 \cdot 5H_2O$ in SDW q $\cdot$ s 500 ml (fehling Sol A)
2. 120 g KOH & 173 g Na-K Tartarate in water & qs 500 ml (fehling Sol B)
3. Mix Fehling Sol. A & B at 1:1 immediately prior to use (Sol. C)

Glass/Plastic wares : Test tube, beaker, glass pipette, test tube stand

Equipments: Balance

Principle

Fehling's test is a specific & highly sensitive test for detection of reducing sugars. Formation of yellow or red ppt. of cuprous oxide denotes the presence of reducing sugars, Rochelle salt acts as the chelating agent in this reaction.

Reaction

$$CuSO_4 + 2\ KOH \rightarrow Cu\ (OH)_2 + K_2SO_4$$
$$Cu(OH)_2 \rightarrow CuO + H_2O$$

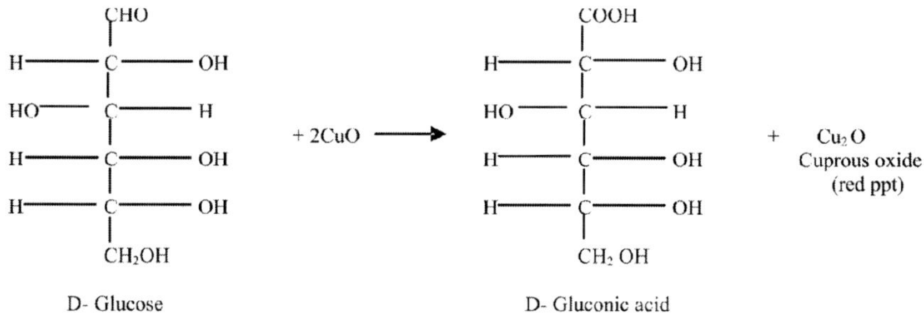

Procedure

1. Take 1 ml of test solution & add 1 ml of Sol. C.
2. Mix thoroughly & cook in Water Bath.

Observations

Appearance of red ppt. of Cu_2O indicates presence of reducing sugar.

7. Benedict's Test

Requirements

Reagents

1. 173 g Na-Citrate & 100 g anhydrous Na_2CO_3 in 600 ml SDW (hot) & q.s to 800 ml (Sol. A)
2. 17.3 g $CuSO_4 \cdot 5H_2O$ in 100 ml hot SDW, cool & q.s 100 ml (Sol. B)
3. Add Sol A to Sol B with constant stirring & q.s 1000 ml (Sol. C)

Glass/Plastic wares : Test tube, beaker, glass pipette, test tube stand

Equipments: Balance

Principle

Benedict's test is more convenient & this reagent is more stable. In this method, sodium citrate functions as a chelating agent. Presence of reducing sugars results in the formation of red ppt. of cuprous oxide.

Reaction

$Na_2CO_3 + 2\ H_2O \rightarrow 2\ NaOH + H_2CO_3$

$2\ NaOH + CuSO_4 \rightarrow Cu\ (OH)_2 + Na_2SO_4$

$Cu\ (OH)_2 \rightarrow CuO + H_2O$

D-Glucose + 2 CuO → D-Gluconic acid + Cu_2O (red ppt)

Procedure

1. Take 1 ml of test solution & add 2 ml of sol. C.
2. Mix thoroughly & cook in Water Bath.

Observation

Appearance of red ppt indicates presence of carbohydrates.

8. Picric Acid Test

Requirements

Reagents

1. 13% Picric acid (w/v) in SDW boil & then cool (Sol A)
2. 10% Na_2CO_3 (Sol B)

Glass/Plastic wares : Test tube, beaker, glass pipette, test tube stand

Equipments: Balance

Principle

It is another test for detection of reducing sugars. The reducing sugars react with picric acid to form a red coloured picramic acid.

Reaction

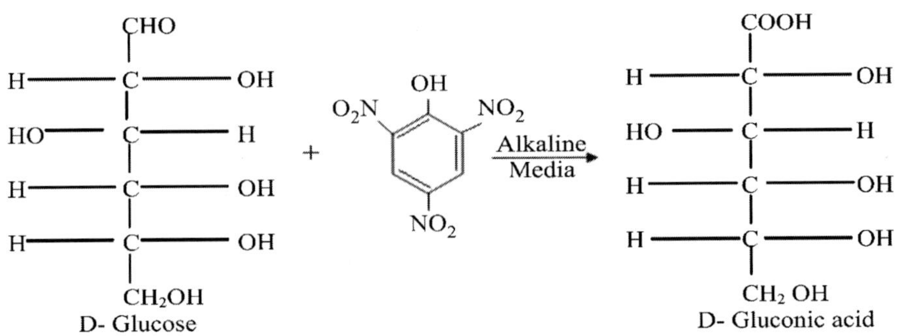

Procedure

1. Take 1 ml test solution and add 1 ml of solution A & 0.5 ml of solution B.
2. Cook in Water Bath & observe the colour.

Observation

Appearance of red colour gives positive test for carbohydrates.

9. Bial's Test

Requirements

Reagents

1. 1.5 g orcinol in 100 ml Conc. HCl (Sol A)
2. 10% $FeCl_3$ Sol (Sol B)
3. Add 20-30 drops of Sol B into Sol A (Sol C)

Glass/Plastic wares : Test tube, beaker, glass pipette, test tube stand

Equipments : Balance

Principle

This test is useful in the determination of pentose sugars; reaction is due to formation of furfural in the acid medium which condenses with orcinol in presence of ferric ions to give a Blue-green coloured complex which is soluble in Butyl alcohol.

Reaction

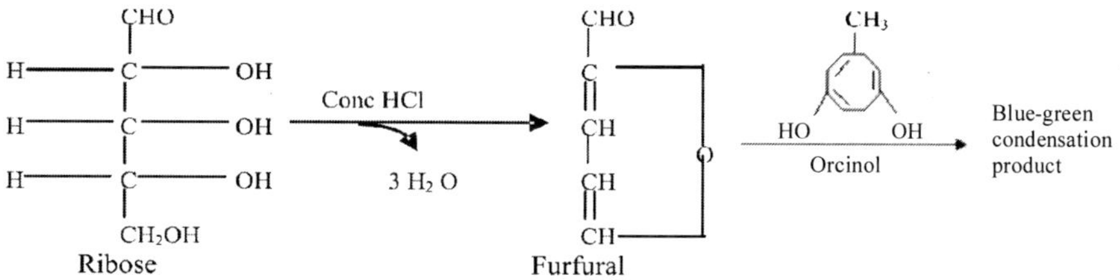

Procedure

1. Take 1 ml of test solution & add 2 ml of solution C.
2. Cook in Water Bath.

Observation

Formation of bluish green colour gives positive test.

Observations

Test Solution Test	Glucose	Xylose	Sucrose	Ribose	Lactose
MOLISCH TEST	+Ve	+Ve	+Ve	+Ve	+Ve
ANTHRONE TEST	+Ve	+Ve	+Ve	+Ve	+Ve
IODINE TEST	+Ve	+Ve	+Ve	+Ve	+Ve
BARFOED'S TEST	+Ve	+Ve	-Ve	+Ve	-Ve
SELIWANOFF'S TEST	-Ve	-Ve	+Ve	-Ve	-Ve
FEHLING'S TEST	+Ve	+Ve	-Ve	+Ve	+Ve
BENEDICT'S TEST	+Ve	+Ve	+Ve	+Ve	+Ve
PICRIC ACID TEST	+Ve	+Ve	+Ve	+Ve	+Ve
BIAL'S TEST	+Ve	+Ve	+Ve	+Ve	+Ve

Experiment – 2 : Estimation of Reducing Sugar by Dinitrosalicylate (DNS) and Plotting Standard Curve by Dextrose

Principle

Several reagents have been implied to assay sugars by reducing their sugar properties. One such compound is 3,5-dinitrosalicyclic acids or DNS. Reducing sugar converts DNS under alkaline conditions to amino nitrosalicyclate, a orange-yellowish compound that has an absorption maxima at 540 nm. The chemistry of the reaction is complicated since standard curves do not always go through the origin and the different sugar gives different colour aids. The method is different colour aids. The method is therefore is not suitable for the detection of complex mixture of reducing sugar.

Requirements

Reagents : Distilled water, DNS, NaOH, Na-K Tartarate, Reducing sugar (Dextrose)

Glass/Plastic wares : Test tube, beaker, glass pipette, test tube stand, measuring cylinder.

Equipments : Balance

Chemical Reactions

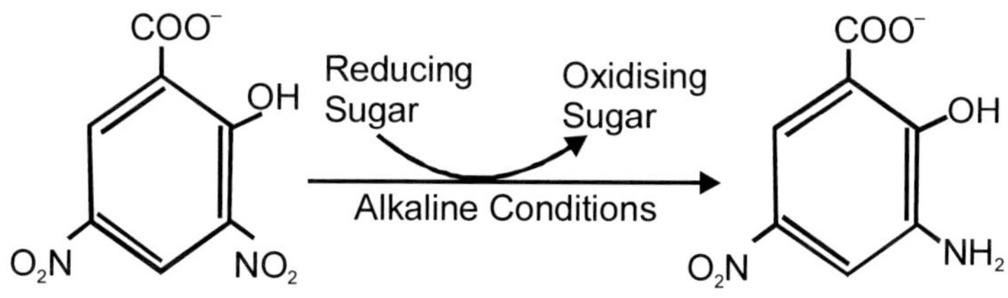

DNS 3- Amino, 5-nitrosalicyclate

Preparation of Reagents

(i) **DNS Reagent:-** 30 gm of Na-K Tartarate weighed accurately and dissolved in 30 ml of SDW, 1 gm of DNS is added to 20 ml of 2 N NaOH and this mixture is transferred to 30 ml of sodium-potassium tartarate. Volume is made upto 100 ml. It is kept in a Boiling water bath to get a sparking clear yellowish orange solution.

(ii) **Stock Solution:-** 0.1 gm of D-Glucose in 100 ml of single distilled water.

Procedure

1. 10 aliquots have been prepared with various gradual elilutions of the stock solution.
2. DNS is added to each tube at ratio of 1:1.
3. The open end of the test tube is wrapped with aluminium foil in order to avoid, moisture entrapment.
4. The tubes are placed in a boiling water bath (temp 80-100^{0}C) and cooked for 5 minutes.
5. Tubes are cooled at room temperature & optical density was

   γ_{540}.
6. Standard curve has been platted by taking conc. in X-axis and γ_{540} in Y-axis and linear regression has been calculated by computing.

Observations

Tubes	Stock Concentration	Stock Volume (μe)	Sdw (μl)	% Conc.	O.d. At Λ_{540}
S_1	0.1 g/dl	110	890	0.011	0.203
S_2	0.1 g/dl	120	880	0.012	0.204
S_3	0.1 g/dl	130	870	0.013	0.215
S_4	0.1 g/dl	140	860	0.014	0.222
S_5	0.1 g/dl	150	850	0.015	0.245
S_6	0.1 g/dl	160	840	0.016	0.265
S_7	0.1 g/dl	170	830	0.017	0.288
S_8	0.1 g/dl	180	820	0.018	0.294
S_9	0.1 g/dl	190	810	0.019	0.317
S_{10}	0.1 g/dl	200	800	0.020	0.336

Experiment – 3 : Estimation of Sugars by Orcinol Method

Principle

Ribose sugar gets converted into furfural in the presence of hot acid orcinol reacts with furfural to give a green coloured complex in the presence of $FeCl_3$ which can be spectrophotometrically measured at 660 nm.

Reaction

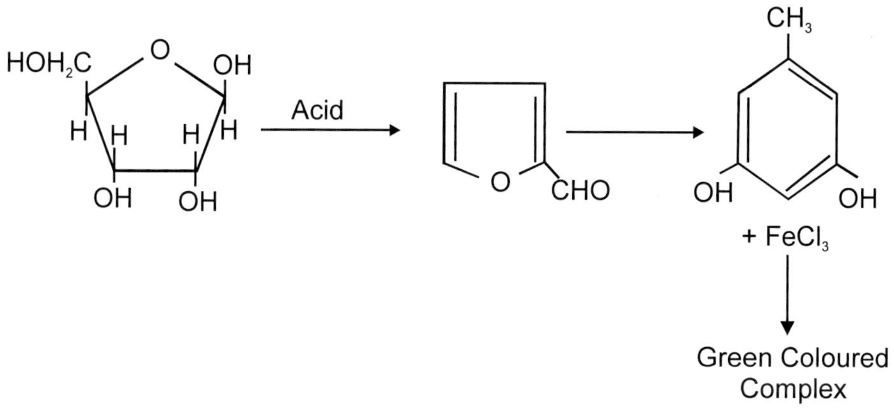

Requirements

Reagents : Distilled water

1. Standard RNA solution: (100 µg/ml) in 0.15 M sodium acetate buffer of pH = 7.2
2. Orcinol acid reagent: 1 ml 10% w/v $FeCl_3$ in 200 ml conc. HCl
3. Orcinol alcoholic solution: (3.5 ml) 6% w/v orcinol in Ethanol

Glass/Plastic wares: test tubes, beakers, measuring cylinder

Equipments: water bath, spectrophotometer

Procedure

1. Take ten test tubes, standard RNA solution is added in various amounts ranging from 0.1-1 ml.
2. Volume is made up by adding 3 ml distilled water.
3. Then 6 ml orcinol acid reagent & 0.4 ml of orcinol alcoholic solution is added to all the test tubes.

4. The tubes are kept in water bath for 15-20 min. at 50-60^{0}C.
5. It is then cooled at room temp. and O.D. is taken at 660 nm with blank as reference.

Observations

Test Tube No.	Stock Conc.	Stock Vol/Ml	Conc. (μg/Ml)	Double Distilled H_{20} (Ml)	Orcinol Acid Reagent	Orcinol Alcohol Solution	O.d At 660 Nm
Blank	100 μg/ml	0.0		3.0	6 ml	0.4 ml	0.009
1.	100 μg/ml	0.1	10	2.9	6 ml	0.4 ml	0.142
2.	100 μg/ml	0.2	20	2.8	6 ml	0.4 ml	0.247
3.	100 μg/ml	0.3	30	2.7	6 ml	0.4 ml	0.33
4.	100 μg/ml	0.4	40	2.6	6 ml	0.4 ml	0.389
5.	100 μg/ml	0.5	50	2.5	6 ml	0.4 ml	0.43
6.	100 μg/ml	0.6	60	2.4	6 ml	0.4 ml	0.47
7.	100 μg/ml	0.7	70	2.3	6 ml	0.4 ml	0.53
8.	100 μg/ml	0.8	80	2.2	6 ml	0.4 ml	0.589

Experiment – 4 : Estimation of Reducing Sugars by Nelson-Somogyi's Method

Principle

The reducing sugars when heated with alkaline copper tartrate reduce the copper from the cupric to cuprous state and thus cuprous oxide is formed. When cuprous oxide is treated with arsenomolybdic acid, the reduction of molybdic acid to molybdenum blue takes place. The blue colour developed is compared with a set of standards in a colorimeter at 620 nm.

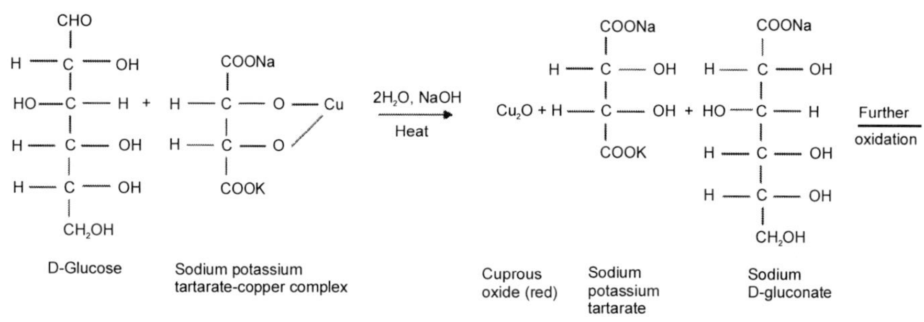

Requirements

Reagents

1. Alkaline copper tartrate : Dissolve 2.5 g anhydrous sodium carbonate, 2g sodium biacarbonate, 2.5g potassium sodium tartrate and 20g anhydrous sodium sulphate in 80 ml water and make up to 10 ml. (Solution A). Dissolve 15g copper sulphate in a small volume of distilled water. Add one drop of sulphuric acid and make up to 100 ml. (Solution B). Mix 4ml of (Solution A) and 96ml of solution (Solution B) before use.
2. Arsenomolybdate reagent: Dissolve 2.5g ammonium molybdate in 45ml water. And 2.5ml sulphuric acid and mix well. Then add 0.3g disodium hydrogen arsenate dissolved in 25ml water. Mix well and incubate at 37^{0}C for 24 to 48h.
3. Standard stock glucose solution: 100ml distilled water.
4. Working standard: Dilute 10ml of stock solution to 100ml with distilled water (100µg/ml).

Glass/Plastic wares: beakers, test tubes, microtips

Equipments: water bath, colorimeter, micropipette

Procedure

1. Weigh 100mg of the sample and extract the sugars with hot 80% alcohol twice (5ml each time).
2. Collect the supernatant and evaporate on water bath.
3. Add 10ml of water and dissolve the sugars.
4. Pipette out aliquots of 0.1 or 0.2 ml of alcohol-free extract to separate test tubes.
5. Pipette out 0.2, 0.4, 0.6, 0.8 and 1ml of the working standard solution into a series of test tubes.
6. Make up the volume in both sample and standard tubes to 2ml with distilled water.
7. Pipette out 2ml distilled water into a separate tube to serve as a blank.
8. Add 1ml of alkaline copper tartrate reagent to each time.
9. Place the tubes in a boiling water for 10 min.
10. Cool the tubes and add 1ml of arsenomolybdic acid reagent to all the tubes.
11. Make up the volume in each tube to 10ml with water.
12. Read the absorbance of blue colour at 620nm after 10 min.
13. From the graph drawn, calculate the amount of reducing sugars present.

Calculation

$$\text{Reducing sugars in sample (\%)} = \frac{\text{Sugar value from graph } (\mu g)}{\text{Aliquot sample used (0.1 or 0.2 ml)}} \times \frac{\text{Total volume of alcohol free extract (10 ml)}}{\text{Weight of sample (100 mg)}} \times \frac{1}{100}$$

Note :

1. *Although the colour is stable, determine the absorbance at a fixed time after the addition of arsenomolybdate reagent.*

Observations

....

....

....

....

Experiment – 5 : Estimation of total Soluble Sugars by Anthrone Method

Principle

The anthrone reaction is the basis of a rapid and convenient method for the determination of hexoses, aldopentoses and hexuronic acids either free or present in polysaccharides. Carbohydrates are dehydrated by conc. H_2SO_4 to form furfural. Furfural condenses with anthrone (10-keto- 9, 10-dihydro-anthracene) to form a blue-green coloured complex which is measured colorimetrically at 630 nm.

Requirements

Reagents

1. 2.5N HCl
2. Anthrone reagent: Dissolve 200mg anthrone in 100ml of ice cold 95% H_2SO_4. Prepare fresh before use.
3. Standard glucose: Stock-Dissolve 100mg in 100ml water, working standard-10ml of stock diluted to 100ml with distilled water. Store refrigerated after during a few drops of toluene.

Glass/Plastic wares: beakers, test tubes, microtips

Equipments: water bath, centrifuge, refrigerator, colorimeter, micropipette

Procedure

1. Weigh 100mg of the sample into a boiling tube.
2. Hydrolyze by keeping it in a boiling water bath for 3h with 5ml of 2.5N HCl and cool to room temperature.
3. Neutralize it with solid sodium carbonate until the effervescence ceases.
4. Make up the volume to 100ml and centrifuge.
5. Collect the supernatant and take 0.5 and 1ml aliquots for analysis.
6. Prepare the standards by taking 0, 0.2, 0.4, 0.6, 0.8 and 1ml of the working standard. '0' serves as blank.
7. Make up the volume to 1ml in all the tubes including the sample tubes by adding distilled water.

8. Then add 4ml of anthrone reagent.
9. Heat for 8 min in a boiling water bath.
10. Cool rapidly and read the green to dark green colour at 630 nm.
11. Draw a standard graph by plotting concentration of the standard on the X-axis versus absorbance on the Y-axis.
12. From the graph calculate the amount of carbohydrates present in the sample tube.

Calculation

$$\text{Amount of carbohydrate present in sample (\% mg)} = \frac{\text{Sugar value from graph (mg)}}{\text{Aliquot sample used (0.5 or 1 ml)}} \times \frac{\text{Total volume of extract (ml)}}{\text{Weight of sample (mg)}} \times 100$$

Note :

1. *Cool the contents of all the tubes on ice before adding ice cold anthrone reagent.*

Observations

....

....

....

....

Experiment – 6 : Estimation of total Soluble Sugars By Phenol – Sulphuric Acid Method

Principle

Simple sugars, oligosaccharides, polysaccharides and their derivatives give an orange-yellow colour when treated with phenol and conc. H_2SO_4. The reaction is sensitive and the colour is stable.

In hot acidic medium glucose is dehydrated to hydroxymethyl furfural. This forms an orange-yellow coloured product with phenol and has absorption maximum at 490nm.

The basis of this reaction is the differential production: of furfural when pentoses are heated with dilute acids and of hydroxymethly furfural when hexoses are heated with dilutes acids. When either product condenses with phenol, an orange to yellow product is formed.

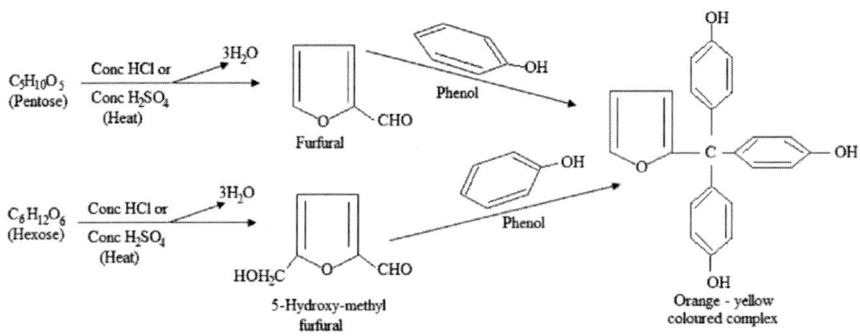

Requirements

Reagents

1. 5% Phenol: Dissolve 50g of redistilled (reagent grade) phenol in water and dilute to one litre.
2. 96% Sulphuric acid (reagent grade).
3. Standard glucose: Stock – 100mg in 100ml of water.

Glass/Plastic wares : beakers, test tubes, microtips

Equipments : water bath, centrifuge, refrigerator, colorimeter, micropipette

Procedure

1. Follow the steps 1 to 4 as given in anthrone method for sample preparation.

2. Pipette out 0.2, 0.4, 0.6, 0.8 and 1ml of working standard into a series of test tubes.
3. Pipette out 0.1 and 0.2 ml of the sample solution in two separate test tubes. Make up the volume in each tube to 1ml with water.
4. Set a blank with 1ml of water.
5. Add 1ml of phenol solution to each tube.
6. Add 5ml of 96% sulphuric acid to each tube and shake well.
7. After 10ml shake the contents in the tubes and place in a water bath at 25-30^{0}C for 20 min.
8. Read the colour at 490nm.
9. Calculate the amount of total carbohydrate present in the sample solution using the standard graph.

Calculation

$$\text{Amount of carbohydrate present in sample (\%)} = \frac{\text{Sugar value from graph } (\mu g)}{\text{Aliquot sample used (0.1 or 0.2 ml)}} \times \frac{\text{Total volume of alcohol free extract (100 ml)}}{\text{Weight of sample (100 mg)}} \times \frac{1}{1000}$$

Observations

....

....

....

....

Experiment – 7 : Estimation of Non-Reducing Sugars

Principle

Non-reducing sugars present in the plant extracts are first hydrolyzed with either sulphuric acid or formic acid to reducing sugars. Then, the total reducing sugars are estimated either by Nelson-Somogyi's or DNS method.

Requirements

Reagents : 1N H_2SO_4, 1N NaOH, 80% alcohol, distilled water

Glass/Plastic wares: beakers, test tubes, microtips

Equipments: micropipette, Methyl red indicator

Procedure

1. Weigh 100mg of the sample and extract the sugars with hot 80% alcohol twice (5ml each time).
2. Collect the supernatant and evaporate on water bath.
3. Add 10ml of water and dissolve the sugars.
4. Pipette out 1ml of extract and add 1ml of 1N H_2SO_4.
5. Hydrolyze the mixture by heating at 49^{0}C for 30 min (the acid hydrolysis is effective in splitting the sucrose-type linkages).
6. Cool the tubes and add 1 or 2 drops of methyl red indicator.
7. Neutralize the contents by adding 1N NaOH dropwise from a pipette. Maintain appropriate reagents blank.
8. Estimate the total reducing sugars by either Nelson-Somogyi's or DNS method as described earlier.

Calculation

$$\text{Non-reducing sugars (\% mg)} = \frac{\text{Sugar value from graph } (\mu g)}{\text{Aliquot sample used (1 ml)}} \times \frac{\text{Total volume of extract (10 ml)}}{\text{Weight of sample (100 mg)}} \times \frac{1}{1000}$$

Note :

1. *The content of non-reducing sugars can also be calculated by subtracting the reducing sugars from total carbohydrate content.*

Observations

....

....

....

....

Experiment – 8 : Estimation of Starch by Anthorne Reagent

Principle

The sample is treated with 80% alcohol to remove sugars and then starch is extracted with perchloric acid. In hot acidic medium starch is hydrolyzed to glucose and dehydrated to hydroxymethyl furfural. This compound forms a green coloured product with anthrone.

Requirements

Reagents : Anthrone : Dissolve 200mg anthrone in 100ml of ice-cold 95% sulphuric acid, 80% Ethanol, 52% Perchloric acid, Standard glucose-10ml of stock diluted to 100ml with water (100 µg/ml).

Glass/Plastic wares: beakers, test tubes, microtips

Equipments: micropipette, centrifuge, refrigerator, colorimeter, water bath, weighing balance

Procedure

1. Homogenize 0.1 to 0.5g of the sample in hot 80% ethanol to remove sugars. Centrifuge and retain the residue. Wash the residue repeatedly with hot 80% ethanol till the washings do not give colour with anthrone reagent. Dry the residue well over a water bath.
2. To the residue add 5.0ml of water and 6.5ml of 52% perchloric acid.
3. Extract at 0^0C for 20 min. centrifuge and save the supernatant.
4. Repeat the extraction using fresh perchloric acid. Centrifuge and pool the supernatants and make up to 100ml.
5. Pipette out 0.1 or 0.2 ml of the supernatant and make up to the volume to 1ml with water.
6. Prepare the standards by taking 0.2, 0.4, 0.6, 0.8 and 1ml of the working standard and make up the volume to 1ml in each tube with water.
7. Add 4ml of anthrone reagent to each tube.
8. Heat for 8 min in a boiling water bath.

9. Cool rapidly and read the intensity of green to dark green colour at 630nm.

Calculation

Find out the glucose content in the sample using the standard graph. Multiply the value by a factor 0.9 to arrive at the starch content.

Observations

....

....

....

....

Experiment – 9 : Determination of Amylose

Principle

The iodine is absorbed within the helical coils of amylose to produce a blue-coloured complex which is measured colorimetically at 590nm.

Requirements

Reagents

1. 1N NaOH
2. 0.1% Phenolphthalein indicator
3. Distilled water
4. Iodine reagent : Dissolve 1g of iodine and 10g of Kl in water and make up to 500ml.
5. Standard amylose : 100mg amylose in 10ml of 1N NaOH and make up to 100ml with water (1mg/ml).

Glass/Plastic wares : beakers, test tubes, microtips

Equipments: micropipette, centrifuge, refrigerator, colorimeter, water bath, weighing balance

Procedure

1. Take 100mg of powdered sample add 1ml of distilled ethanol, mix well then add 10ml of 1N NaOH. Leave it overnight and make up the volume to 100ml.
2. To 2.5ml of extract, add about 20ml of water, three drops of phenolphthalein indicator and mix.
3. Add 0.1N HCl drop by drop until the pink colour just disappears.
4. Add add 1ml of iodine reagent and make up the volume to 50ml and read the absorbance at 590nm.
5. Calculate the amount of amylose in the sample using the standard curve prepared from aylose (range 0.2-1mg) against a blank for which dilute 1ml of iodine reagent to 50ml with water.

Calculate

Absorbance corresponds to 2.5ml of test solution = 'x' mg amylose in test solution.

$$\therefore \text{100ml contains} = \therefore \frac{x}{2.5} \times 100 = \text{\% amylose}$$

Note :

1. *The sample suspension may be heated for 10 min in a boiling water bath instead of overnight dissolution.*
2. *The amount of amylopectin is obtained by subtracting the amylose content from that of starch.*

Observations

....

....

....

....

Experiment – 10 : Estimation of Maltose from any Fruit Juice

Principle

The maltose content is estimated using 3, 5-dinitrosalicylate which forms an orange-red colour with maltose as a reducing sugar which can be measured colorimetrically at 520nm.

Requirements

Reagents

1. Dinitrosalicylic acid reagent (DNS reagent) : Dissolve 1g of 3, 5-dinitrosalicylate, 30g of sodium potassium tartrate and 1.6g of NaOH in water and make up to 100ml.
2. Standard maltose solution : 1mg/ml water.

Glass/Plastic wares : beakers, test tubes, microtips

Equipments : micropipette, centrifuge, refrigerator, colorimeter, water bath, weighing balance

Procedure

1. Weigh 100mg of sample and extract maltose with 80% warm alcohol twice.
2. Collect the supernatant and evaporate a known volume (2-3 ml) of alcohol extract on a hot water bath. Note the volume.
3. Make up the volume to 2ml with water.
4. Add 2ml of DNS reagent, mix, cover the tube with marble and keep in boiling water bath for 10 min.
5. Cool the dilute to 10ml with water.
6. Measure the absorbance of orange-red colour at 520nm (green filter) against a reagent blank.
7. Calculate the amount of maltose in sample using standard graph prepared from maltose (range 0.1-2mg) similarly.

Calculation

$$\text{Amylose content (\%)} = \frac{\text{Sugar value from graph } (\mu g)}{\text{Aliquot sample used (2 or 3 ml)}} \times \frac{\text{Total volume of extract (10 ml)}}{\text{Weight of sample (100 mg)}}$$

Observations

....

....

....

....

Experiment – 11 : Estimation of Cellulose from Fresh Plants

Principle

Cellulose undergoes acetolysis with acetic/nitric reagent to form acetylated cellodextrins which get dissolved and hydrolyzed to form glucose units on treatment with 67% H_2SO_4. On dehydration with H_2SO_4, glucose forms 5-hydroxymethyl furfural which on reaction with anthrone gives a green coloured product. The colour intensity can be measured at 630nm.

Requirements

Reagents

1. Acetic/nitric reagent: Mix 150ml of 80% acetic acid with 15ml of conc. HNO_3.
2. Anthrone reagent: 200mg anthrone/100ml of conc. H_2SO_4 (prepare fresh and chill before use).
3. 67% H_2SO_4.
4. Standard cellulose solution: To 100mg of cellulose, add 10ml of 67% H_2SO_4 and leave for 1h. Dilute 1ml of the solution to 100ml (100µg/ml).

Glass/Plastic wares: beakers, test tubes, microtips

Equipments: micropipette, centrifuge, refrigerator, colorimeter, water bath, weighing balance

Procedure

1. Take about 0.5-1.0g of sample, add 3ml of acetic: nitric reagent and mix using a vortex mixer.
2. Place in a water bath at 100^{0}C for 30 min.
3. Cool and centrifuge for 15-20 min. Discard the supernatant.
4. Wash the residue with water, add 10ml of 67% H_2SO_4 and leave it for 1h.
5. Dilute 1ml of this solution to 100ml. To 1ml of the diluted solution, add 10ml of anthrone reagent and mix well.
6. Heat the tubes in a boiling water bath for 10 min., cool and measure the absorbance at 630nm.
7. Prepare the blank with anthrone reagent and water.

8. Prepare standard curve by taking 0.4 to 2ml of standard cellulose solution (corresponding to 4-200 µg of cellulose), equalize the volume and add anthrone reagent and develop the colour as above.
9. Calculate the amount of cellulose in the sample from the standard graph.

Observations

....

....

....

....

Experiment – 12 : Estimation of Hemicellose

Principle

The sample is refluxed with neutral detergent solution to remove the water-solubles and materials other than the fibrous component. The left out material is weighed after filtration and expressed as neutral detergent fibre (NDF).

Requirements

Reagents

1. Acetone
2. Sodium sulphite
3. Decahydronaphthalene
4. Neutral detergent solution: Dissolve 18.61g of disodium ethylene-diaminetetraacetate and 6.81g of sodium borate decahydrate in about 200ml of water by heating. To this, add about 100-200ml of a solution containing 30g of sodium lauryl sulphate and 10ml of 2-ethoxy ethanol. Then add about 100ml of a solution containing 4.5g of disodium hydrogen phosphate. Adjust the pH to 7.0 and make up to one litre.

Glass/Plastic wares: beakers, test tubes, microtips

Equipments: micropipette, centrifuge, refrigerator, colorimeter, water bath, weighing balance

Procedure

1. Take 1g of the powdered sample in a refluxing flask add 10ml of cold neutral detergent solution.
2. Add 2ml of decahydronaphthalene and 0.5g of sodium sulphite.
3. Heat to boiling and reflux for 60 min.
4. Filter through sintered glass crucible (G-2) by suction and wash with hot water.
5. Wash twice with acetone, transfer the residue to a crucible and dry at 100^{0}C for 8h.
6. Cool the crucible in a desiccator and weigh.

Calculation

$$\text{Hemicelluloses} = \underset{\text{fibre (NDF)}}{\text{Neutral detergent}} - \underset{\text{fibre (ADF)}}{\text{Acid detergent}}$$

Note :

See under 'Lignin' for determining 'acid detergent fibre'.

Observations

....

....

....

....

Experiment – 13 : Estimation of Crude Fibre

Principle

During the acid and subsequent alkali treatment, oxidative hydrolytic degradation of the native cellulose and considerable degradation of lignin occur. The residue obtained after final filtration is weighted, incinerated, cooled and weighed again. The loss in weight gives the crude fibre content.

Requirements

Reagents

1. 0.255 ± 0.005 N Standard H_2SO_4
2. 0.313 ± 0.005 N Standard NaOH

Glass/Plastic wares : beakers, test tubes, microtips

Equipments : micropipette, centrifuge, refrigerator, colorimeter, water bath, weighing balance

Procedure

1. Extract 2g of ground sample with ether or petroleum ether to remove fat (initial boiling temperature 35-38°C and final temperature, 52°C), if fat content is less than 1%, extraction may be omitted.
2. Boil 2g of dried sample with 200ml of H_2SO_4 for 30 min with bumping chips.
3. Filter through muslin cloth and wash with boiling water until washings are free of acid.
4. Boil the residue with 200ml of NaOH for 30 min.
5. Filter through muslin cloth again and wash with 25 ml of boiling H_2SO_4, three 50ml portions of water and 25ml of alcohol.
6. Remove the residue and transfer to pre-weighed ashing dish ('W_1', g).
7. Dry the residue for 2h at 130 ± 2^0C, cool) in a desiccator and weigh ('W_2', g).
8. Ignite for 30 min at 600 ± 15^0C.
9. Cool in a desiccator and reweigh ('W_3', g).

Calculation

Loss in weight x $(W_2 - W_1) - (W_3 - W_1)$

$$\% \text{ Crude fibre content} = \frac{\text{on ignition}}{\text{Weight of sample (g)}} \times 100$$

Observations

....

....

....

....

Additional Reading

Ashford, D.A. (1992). Carbohydrate analysis. Curr. Opin. Biotechnol. 3(1): 45.

Biermann, C.J. and McGinnis, G.D. (editors) (1989). Analysis of carbohydrates by GLC and MS. CRC Press Inc., Boca Raton, FL.

Binkly, R.W. (1988). Modern Carbohydrate Chemistry. Marcel Dekker, Inc., San Diego, CA.

Burchfield M Pand Storrs EE(1962). Biochemical Application of Gas Chromatography Academic Press, New York.

Kleinev. I.S. and Dotti L.B. (1966). Laboratory Instructions in Biochemistry. The C.V. Mosby Company. St. Lonis U.S.A.

Whelan, W.J. (1975). Biochemistry of Carbohydrates. Butterworth, London.

Whistler, R.L. and Wolfrom, M.L. (editors) (1962-1980, 8 volume). Methods in Carbohydrate Chemistry. Academic Press Inc., New York.

❑❑❑

Unit -3

Amino Acids and Proteins

Amino acids are essential components of all living cells as building blocks of proteins. They are also metabolically active and supply substrates for many other biochemical reactions. In a bacterial cell, about 0.4% of the dry weight is amino acids. The amino units in other living systems varies according to species, age and part of the body. All proteins are macromolecules because of their very high molecular weights. These are the polymers i.e. chain like molecules produced by joining a number of small units of amino acids called monomers. The amino acids are therefore regarded as building blocks of proteins.

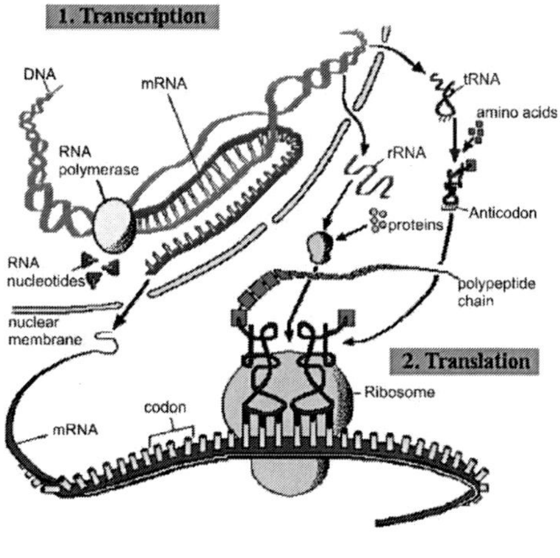

Fig. 3.1 : Protein synthesis process.

Chemically amino acids are carboxylic acids containing at least one amino group. In protein amino acids it is always the α- carbon atom, which is animated and hence they are called α- amino acids. The general structural formula of the amino acids is

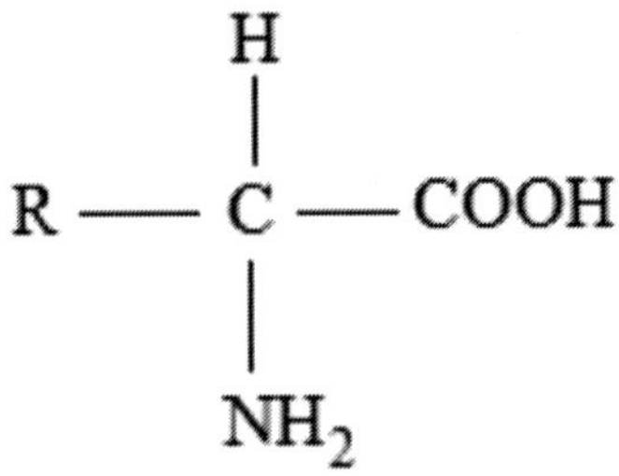

Where R may be –H as in glycine or $-CH_3$ as in alanine or some other group in some other amino acid.

Although over 100 amino acids have been shown to be present in various plants and animals only 20 of them (L-isomers) are formed as constituent of most proteins. These 20 amino acids of proteins are often referred to as standard, primary or normal amino acids, to distinguish them from others. The almost universal use of these 20 amino acids for the synthesis of various protein molecules is "one of nature's enigmatic rules".

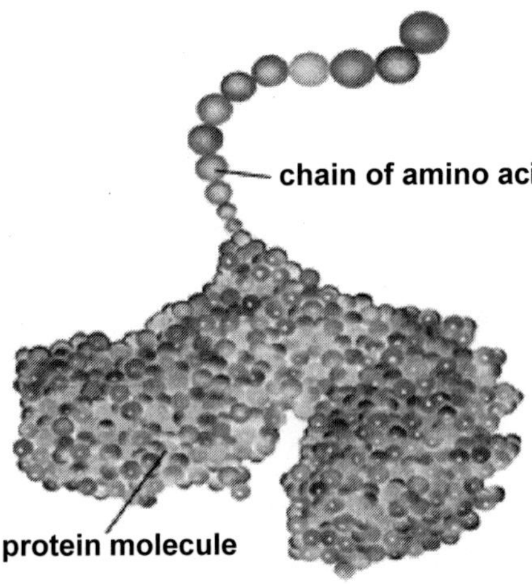

The term protein (proteios[G] = pre – eminent of first) was suggested by Swedish Chemist Berzelius and used by GJ Mulder for the first time in his text book in 1840. All the basic functions of life depend upon specific proteins. Indeed we do not know of any life without proteins. They are present in each cell in one from or the other. Proteins account on average about one fourth to one third of the total dry weight of any living organism. Proteins take part in the structural organization of several cellular components. It has been known for a long time that the proteins are high molecular weight compounds ranging from about 6000 to many million and made of several amino-acid.

- ***"An inventor is a person who makes an ingenious arrangement of wheels, leavers and springs, and believes it civilization."***

Ambrose Bierce

Experiment – 1 : Identification of N-terminal Amino Acid of a Protein using Dansyl Chloride Method

Principle

Crystals of amino acids are white and its solution appears colourless. Identification of acids depends on the amount of carbon, oxygen, nitrogen, sulfur, its solubility, melting point, colligative properties, etc. Specifically, dansyl chloride reacts with the free amino group of the *N*-terminal acid residues of a protein. Upon acid hydrolysis it results in formation of dansylated amino acid the other free amino acids are separated. They can be identified using ultraviolet light. Under light dansyl amino acid appears greenish-yellow in colour.

Requirements

Reagent : Dansyl chloride solution (2.5 mg/ml in acetone) (prepare fresh) and keep at 4^0C refrigerator, 0.2M aqueous solution of sodium bicarbonate (keep at 4^0C, when not in use). Prepare two chromatography solvents as below:

Solvent 1: Formic acid: water (4.5: 100, (v/v)

Solvent 2: Toluene: acetic acid (8:1, v/v)

Glass/Plastic wares: Ignition tubes/test tubes referred to as 'dansyl tubes' (5 cm x 2 mm) Dansyl plates [polyamide thin layer plates (5 cm x 5 cm coated both sides)] coated both sides. Number the plates with a pencil in the top corner.

Mark the origin for loading with a pencil 0.3 – 0.5 cm in from each edge in the lower left hand corner of the numbered side of the plate (the origin for loading on the reverse on the plate should be immediately behind the loading position for the plate with a pencil 0.3 - 0.5 mm from each edge in the lower right hand corner).

Equipments: Ultraviolet source (254 μm or 265 μm), vacuum pump, desiccators, refrigerator.

Procedure

A. Amino Acids

(i) Prepare the stock solution (1 mg/ml) of standard dansyl amino acid in 0.2M NaOH (standard dansyl amino acid are – Pro,

Leu, Phe, Thr, Glu, Arg. Store the stock solution 4^0C when not in use).

(ii) Transfer 10 µl of each amino acid solution in a dansyl tube and mix gently. Add 30 dansyl chloride solution (5 µl for each amino acid).

(iii) Add 10 µl of unknown amino acid and 5 µl of dansyl chloride in a separate dansyl tube.

(iv) Seal the tube with parafilm and incubate at 37^0 C for 1 hour at room temperature for 3 hrs.

(v) Dry the samples in vacuum because the small volume of liquid present will take several minutes.

(vi) Dissolve the dried samples in 10 µl of acetone using a capillary tube. Load 1 µl aliquot the standard mixture at the origin (the diameter of the spot should not be allowed to exceed 1-2 mm).

(vii) Load 1 µl of unknown amino acid on the reverse side.

(viii) Place the plate in the first chromatography solvent when the loaded sample is fully spread. Allow the plate to develop and the solvent to reaches the top. Dry both sides of the plate.

(ix) Observe the plate under UV light.

B. For Proteins

(i) Transfer 10 µg of protein in 10 µl of 0.2 M $NaHCO_3$ in 1 ml stoppered tube.

(ii) Pour 10 µl of dansyl chloride solution in to it.

(iii) Stopper the tube and incubate at 37^0C for 1 hour (or at room temperature for 3 hours).

(iv) Dry the sample in vacuum.

(v) Transfer 20 µl of 6 N HCl to 1 ml tube and close the tube tightly with stopper. Hydrolyse it keeping in an oven maintained at 105 ± 1^0C for 24 hours.

(vi) Dry the samples in vacuum.

(vii) Dissolve the dried sample in 10 µl of acetone and load 1 µl aliquots of the standard mixture at the origin using a fine capillary tube. Diameter of the spot should not be allowed to exceed 1-2 mm.

(viii) Load 1 µl of protein hydrolysate containing the unknown dansyl amino acids on the reverse side.

(ix) Place the plate in the first chromatography solvent when the loaded samples are completely dry. Allow the plate to develop and reach the solvent to the top. Dry both sides of the plate.

(x) Develop the dansyl plate in the second solvent at right angle to the direction of development as found in the first solvent. Wait to reach the solvent to the top, and dry the plate.

(xi) Observe the plate under UV light.

Observations

You should identify three major fluorescent areas at the bottom of the plate. Dansyl chloride is hydrolysed into dansyl hydroxide. The hydrolysed product appears as blue fluorescent area. Dansyl amide (produced by side reaction with NIS_3) shows blue-green fluorescence and is about one third of the way up plate. These two spots are useful internal marker which appears on all plates. The third sport fluorescence green and will correspond to the dansyl derivative of the amino acid. Solvent 2 separates the dansyl derivatives of hydrophobic amino acids and some natural proteins.

Experiment – 2 : Isolation of Milk Protein (Casein) from the Milk

Principle

Proteins are present inside as well as outside the cells of all the living organisms. They do may function inside the cells occurring as enzymes, hormones, etc. Most of them contain 20 amino acids. Proteins are the building blocks of all cells, cellular components and give structural stability.

Caseins are the main group of milk protein. Further several proteins can be resolved from this group such as a_s, b, g and k on the basis of electrophoresis mobility under an electric field. It is selectively precipitated at pH 4.6. At this pH, caseins have a net charge zero because they do not move under an electric field. It explains that the number of positive charges is equal to the number of negative charges at this pH. This is called isoelectric point. Besides minimum solubility of any zwitter-ionic molecule (those having both a negative and positive charge generating residues within a molecule) is found at its isoelectric point. Such molecules may be separated by filtration.

There are many methods for the estimation of proteins in the sample (milk) e.g. lowery's methods and Bradford methods. In this method, isoelectric precipitation principle is used for the isolation of casein from milk.

Requirements

Skimmed milk sample (centrifuge 50 ml milk at 2,000 rpm for 10 minute, remove fat layer and use the solution for the test).

Reagent: 1N HCl (100 ml); 0.4M HCl (90.7 ml of stock solution and diluted to 2 litre); 0.4M NaOH

Glass/Plastic wares: Beaker, dropper, glass rod, measuring cylinder, test tubes

Equipments: pH meter, cheese cloth, Water bath, oven, refrigerator, centrifuge, thermometer, balance

Procedure

1. Take 50 ml skimmed milk sample and dilute it by adding 50 ml water.
2. Keep this mixture in a water bath at 30^0C.

3. By using a dropper and 0.4M HCl in the mixture drop by drop stirring continuously.
4. Check the pH continuously by using a pH meter until pH 4.6 is obtained.
5. Filter the content through cheese cloths when mixture attains the pH 4.6.
6. Collect the precipitated casein curd and measure the weight of it.
7. Take known amount of wet sample and dry it in an oven at 100^0C.
8. Check the weight of dried sample continuously till it attains a constant weight.

Observations

Record the results as below-

Amount of 0.4M HCl added (ml)	pH
	4.6

Volume of the milk used	= …………..
Temperature of the milk	= …………..
Initial pH of the milk	= …………..
(Amount of 0.4M HCl was used (ml), till the pH raised to 4.6 (used volume)	
Wet weight of the sample	= ………….. gram after drying at 100^0C
Dry weight of the sample	= ………….. gram

Experiment – 3 : Estimation of Total Protein from Plant Fresh Tissue

Principle

Proteins are the most important chemicals of living cells. They perform a variety of functions such as building the cell wall and membrane structure, assisting the important of smaller molecules across the membranes acting as storage form of nitrogen in the seeds and catalyzing the biochemical reactions. The catalytic function of proteins is the most important and dynamic aspect of their existence. It is believed that there are no chemical reactions that take place in living cells that are not regulated by one or more enzymic proteins. Proteins are large molecules with molecular weight ranging between 6 and 200 k. Da. All proteins are made up of 20 or less types of amino acids.

Requirements

Reagent : distilled water, Folin – phenol, NaOH, bovine serum albumin (BSA), $Na_2CO_3.H_2O$, $CuSO_4.5H_2O$, $NaKC_4H_4O_6.4H_2O$.

Method for the Preparation of Reagents

Sodium hydroxide solution: For preparation of 0.1 N NaOH solution 4g of NaOH dissolve in one litre of distilled water.

Reagent A: For 2% $Na_2CO_3.H_2O$, 2 g of the salt dissolve in 100 ml of 0.1 n NaOH.

Reagent B: 0.5% copper sulphate ($CuSO_4.5H_2O$) dissolve in 1% sodium potassium tartarate ($NaKC_4H_4O_6.4H_2O$).

Reagent C: One ml of reagent B mix with 50 ml of reagent A.

Glass/Plastic wares: Beaker, funnel, measuring cylinder, petridish, conical flask, test tube.

Equipments: mortar and pestle, filter paper, centrifuge, water bath, spectrophotometer.

Procedure

Total protein in dried material extract with 80% ethanol. The extract keeps as such for 30 min at room temperature (25 – 28^{0}C) and then centrifuge at 12000 x g for 20 min. The clear supernatant use for amino acid determination and the residue for the proteins. The protein measure by the folin – phenol method. For this the residue dissolves in 10 ml of

0.1 N NaOH. Two ml of this solution take and mix with 3 ml of freshly prepared reagent C and keep in a water bath at 25 - 30^{0}C for 15 minutes. Finally 1.0 ml of 1.0 N Folin - phenol reagent adds to it and absorbance of the bluish mixture record at 660 nm. A standard curve prepares using bovine serum albumin (Fig. 3.2), and the amount of protein calculates on dry weight basis.

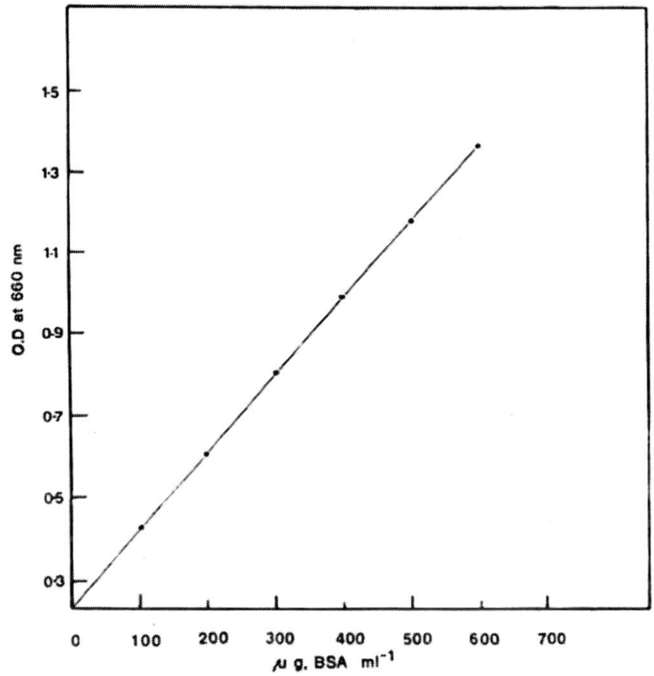

Fig. 3.2 : Calibration curve for protein from bovine serum albumin.

Bovine - Serum Albumin : For preparing standard stock solution 100 mg of BSA dissolve in 100 ml of 0.1 N NaOH.

Observations

....

....

....

....

Experiment – 4 : Estimation of Total Amino Acid from Plant Fresh Tissue

Principle

The amino acids are colourless ionic compounds that form the basic building blocks of proteins. Apart from being bound as proteins, amino acids also exist in the free form in many tissues and are known as free amino acids. They are mostly water soluble in nature. Very often in plants during disease conditions, the free amino acid composition exhibits a change and hence, the measurement of the total free amino acids gives the physiological and health status of the plants.

Ninhydrin, a powerful oxidizing agent, decarboxylates the alpha-amino acids and yields an intensively coloured bluish purple product which is colorimetrically measured at 570 nm.

Ninhydrin + Alpha-amino acid → Hydrindantin + Decarboxylated amino acid + Carbondioxide + Ammonia

Hydrindantin + Ninhydrin + Ammonia → Purple Red Coloured Product + Water

Requirements

Fresh plant tissue,

Regent : distilled water, ninhydrin solution, alcohol,

Glass/Plastic wares : Beaker, funnel, measuring cylinder, petridish, conical flask, test tube, test tube stand,

Equipments : mortar and pestle, filter paper, micropipettes, spectrophotometer.

Procedure

Total amino acid content measure colorimetrically by the method of Yemm and Cocking (1955), supernatant obtain from 80% ethanol extract tissue take in a petridish and keep at a warm plate for evaporation of the ethanol. The residue leave after evaporation dissolve in 5.0 ml of hot distilled water. To 3.0 ml of this solution, 1.0 ml of 0.1% ninhydrin solution prepared in 80% alcohol add and the mixture keep in boiling water bath for 10 min. After cooling, the absorbance measure at 540 nm by spectrophotometer. The amount of total amino acid calculate with the help of a standard curve prepare from lysine (Fig 3.3).

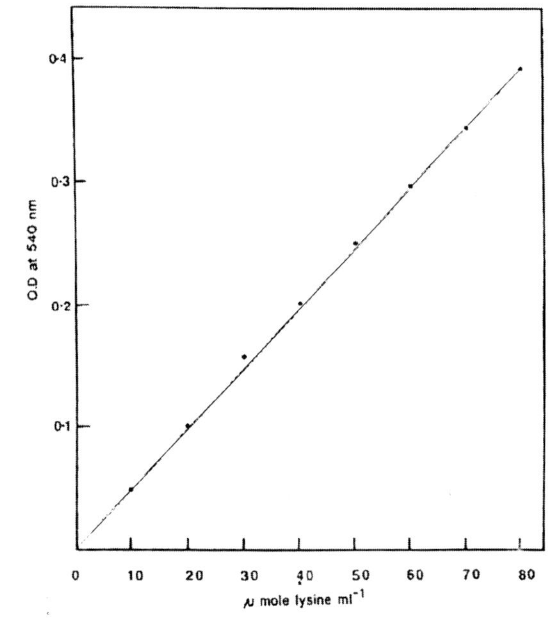

Fig. 3.3 : Calibration curve for lysine.

The reaction of amino acids with ninhydrin is an oxidation reduction reaction. In this reaction amino - acid is oxidized to an aldehyde; CO_2 and NH_3 is released and the reduced ninhydrin then reacts with NH_3 and some more ninhydrin to form purple colour compound, as per following reaction:

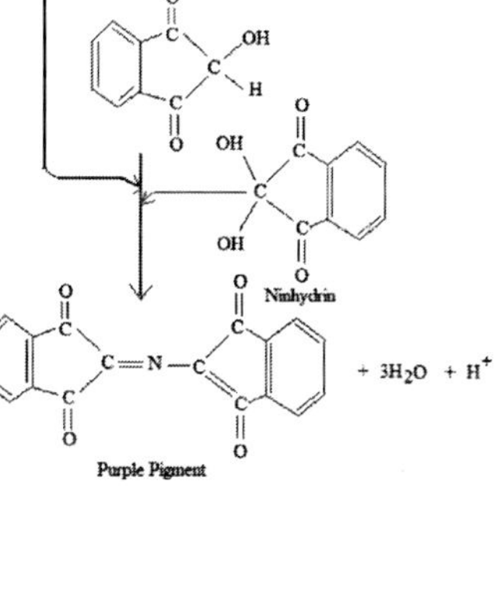

Observations

....

....

....

....

Experiment – 5 : Identify the Amino Acid present in the given sample by Qualitative Method

A. Ninhydrin Test (for Amino Acid)

Principle

This test is due to reaction between α-amino group of free amino acid and ninhydrin (triketohydrindene hydrate). Ninhydrin is a powerful oxidizing agent and in its presence, amino acids undergo oxidative deamination liberating ammonia, carbon dioxide, a corresponding aldehyde and reduced form of ninhydrin. The ammonia formed from α-amino group reacts with ninhydrin and its reduced product (hydrindantin) to give a blue substance diketohydrin (Ruhemann's purple). However, in case of imino acid like proline, a different product having a bright yellow colour is formed. Asparagine which has a free amide group reacts to give a brown coloured product. This test is also given by proteins and peptides.

Reaction

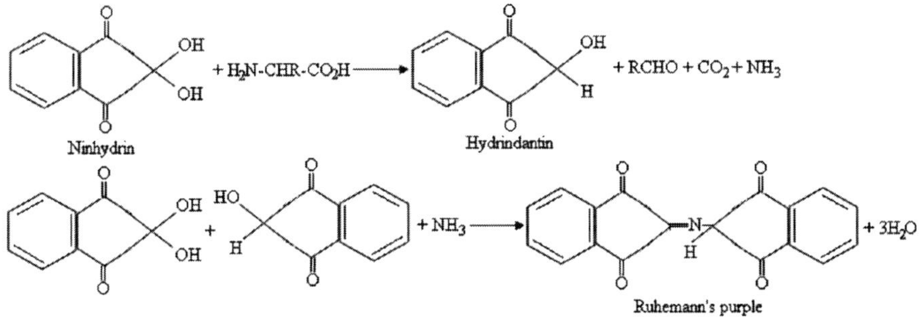

Requirements

Reagents

1. Test solution: prepared solution containing 50 μg/ml of individual amino acids.
2. Ninhydrin Reagent: 0.2% Ninhydrin prepared in acetone.

Equipments: Balance, Water bath.

Glassware : Test tubes, beaker, glass pipette etc.

Procedure

1. Add 2-5 drops of ninhydrin solution to 1 ml of test solution or sample preparation. Mix and keep for 5 min in BWB.

B. Xanthoproteic Test (for Aromatic Amino Acid)

Principle

Aromatic amino acids (tyrosine, tryptophan, and phenylalanine) form yellow nitro derivatives on heating with conc. HNO_3. The salt of these derivatives are orange in colour. Proteins containing these amino acids also give a positive response of this test.

Reaction

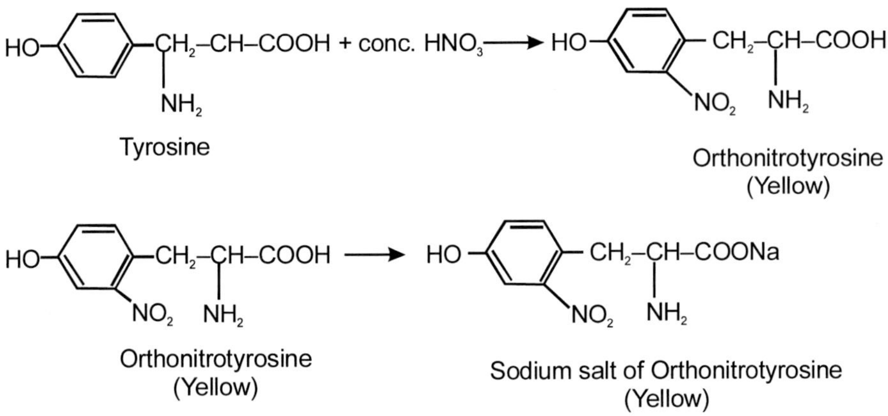

Requirements

Reagents

1. Conc. HNO_3
2. NaOH solution (40%, w/v): Dissolve 40gm of NaOH in water and make the final volume to 100ml.
3. Test solution: prepare separate solutions containing 50µg/ml of amino acids such as tyrosine, glycine, tryptophan, phenylalanine, lysine, etc.

Equipments : Balance, Water bath.

Glassware : Test tubes, beaker, glass pipette etc.

Procedure

1. Add 1ml conc. HNO_3 into 1 ml of test solution.
2. Mix the contents and keep in BWB for 1 min. and allow to cool.
3. Slowly pipette 40% NaOH till the solution becomes alkaline.
4. Note down the change in the color.

C. Millon's Test (for Tyrosine)

Principle

Amino acids are compounds containing hydroxybenzene radical. It react with Million's reagent which constitute of mercurous and mercuric nitrates containing HNO_3 to form a red complex thus this test is specific for tyrosine.

Reaction

$$Hg + 4HNO_3 \rightarrow Hg(NO_3)_2 + 2NO_2 + 2H_2O$$

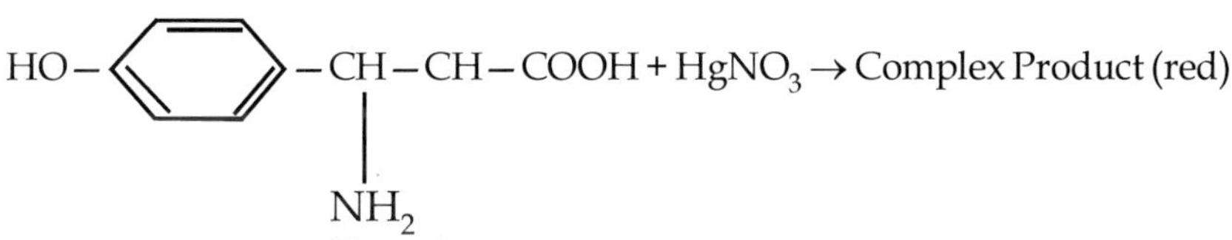

Tyrosine

Requirements

Reagents

1. Million's reagent: Dissolve 10g of mercury (Hg) metal in 20 ml of concentrated HNO_3 dilute to 100 ml with after cessation of evolution of the brown fumes.
2. Test solution of different amino acids made as in the preceding test. Use a solution of phenol also as a test sample.

Equipments : Balance, Water bath.

Glassware : Test tubes, beaker, glass pipette etc.

Procedure

1. Add 2-5 drops of Millon's reagent into 1 ml of test solution.

2. Mix well and place the test tubes in boiling water bath for 5 min.

The same experiment may be carried out with a modified Millon's reagent which contains 10% mercuric sulphate in 10%H_2SO_4. To 1 ml of test solution, add 1 ml of modified millon's reagent and boil gently for 1min. cool under running tap water and the add one or two drops of 1% $NaNO_2$ solution and heat slightly.

Observations

Appearance of red colour denotes presence of tyrosine. The advantage of using modified Million's test lies in its being susceptible to interference from inorganic salts.

Experiment – 6 : Estimate the Concentration of Amino Acids in given sample by Ninhydrin Method

Principle

Ninhydrin is a powerful oxidizing agent and in its presence, amino acids undergo oxidative deamination liberating ammonia, carbon dioxide, a corresponding aldehyde and reduced form of ninhydrin in between the pH 4-8. The ammonia formed from α- amino group reacts with ninhydrin and its reduced product (hydrindantin) to give a blue substance diketohydrin (Ruhemann's purple). However, in case is formed. Asparagine which has a free amide group reacts to give a brown coloured product. This test is also given by proteins and peptides.

Requirements

Reagents :

1. Standard amino acid solution: (Leucine/ aspartic acids/ arginine/ proline: 0.1 mmol/l): Prepare 0.1mM amino acid solution by dissolving gram molecular weight into liter of D/W.
2. Ninhydrin reagent: Dissolve 0.8g of ninhydrin in 25ml acetone and add 25 ml of 0.2 M acetate buffer (pH 5.5). Store in brown bottle to protect from light.
3. 0.2 M acetate buffer (pH 5.5): Prepare 0.2M Sodium acetate and make the pH 5.5 with glacial acetic acid. Make the final volume 1 liter with D/W.
4. 70% ethanol
5. Test sample

Equipments : Balance, pH meter, water bath, spectrophotometer.

Glassware : Test tubes, conical flasks, glass pipette etc.

Range of the method: 10-100 μM

Working concentration: 0.1 mmol/l

Procedure

1. Take clean and dried test tubes. Mark all the tubes as per the protocol table.
2. Take all the reagents along with amino acids (standard) solution with 10, 20, 30, up to 100 μM and test solution sequentially as per the protocol table.

3. Make up the final volume 4ml with D/W along with the blank tubes
4. Add 1ml Ninhydrin reagent in all tubes. Mix all the reagents properly and incubate the mixture in BWB as per the protocol table.
5. After cooling add 1ml of 70% ethanol in all tubes.
6. Mix it properly and measure the color complex spectrophotometrically at 550nm.
7. Draw the standard curve for the Concentration of amino acids vs Optical Density (O.D.)
8. Put the O.D. of the unknown solution and calculate out the concentration of the amino acids present in given sample from standard graph and formula (given below) and express your result in μM of amino acids present in 100ml.

Formula

$$\frac{\text{OD (test)}}{\text{OD (std)}} \times \frac{\text{Conc (std)}}{\text{Aliquot (test)}} \times 100$$

Calculation

1. By direct method

Conc of U1 at 0.190 OD = 0.023 μM/0.5 ml

Conc. of U2 at 0.45 OD = 0.058 μM/1.0 ml

$\therefore$ U1 = 0.046 μM/ml = 4.6 μM/100ml

$\therefore$ U2 = 0.058 μM/ml = 5.8 μM/100ml

$$\because U = \frac{4.6 + 5.8}{2} = 5.2 \ \mu M/100ml$$

2. By slope method

Slope = 0.4/0.05 = 8

$$U1 = \frac{\text{OD (test)}}{\text{Slope}} = 0.19/8 = 0.03 \ \mu M/0.5ml$$

$$= 6 \ \mu M/100ml$$

$$U2 = \frac{\text{OD (test)}}{\text{Slope}} = 0.45/8 = 0.05 \ \mu M/ml$$

$$= 5 \ \mu M/100ml$$

$$U = \frac{6 + 5}{2} = 5.5\ \mu M/100\ ml$$

3. By using formula

U1 = 0.19/0.76 x 0.1/0.5 x 100/1000 = 0.05 μM/ml
= 5 μM/100ml

U2 = 0.45/0.76 x 0.1/1 x 100/1000 = 0.059 μM/ml
= 5.9 μM/100ml

$$U = \frac{5 + 5.9}{2} = 5.45\ \mu M/100ml$$

$$\overline{U} = \frac{5.2 + 5.5 + 5.45}{2} = 5.38\ \mu M/100ml$$

Observation

Concentration of standard = 0.1 mM/ml

Tubes	Aliquote	Conc.	D/W	Ninhydrin reagent (ml)		50% ethanol		OD at 550 nm
B	-	-	4.0	1.0	Water bath for 10′ & then cool	1.0	Incubate the tubes for 10′ at room temperature	
S1	0.2	0.02	3.8	1.0		1.0		0.07
S2	0.4	0.04	3.6	1.0		1.0		0.38
S3	0.6	0.06	3.4	1.0		1.0		0.59
S4	0.8	0.08	3.2	1.0		1.0		0.69
S5	1.0	0.10	3.0	1.0		1.0		0.76
U1	0.5		3.5	1.0		1.0		0.19
U2	1.0		3.0	1.0		1.0		0.45

Result

By performing Ninhydrin method, we can estimate the concentration of amino present in a given unknown sample.

Conclusion/Interpretation

The concentration of amino acids is 5.38 μM/100.

Concentration by:-

(a) By direct method = **5.2 μM/100 ml**

(b) By slope method = **5.5 μM/100 ml**

(c) By using method = **5.45 μM/100 ml**

Experiment – 7 : Estimate the Concentration of Protein in given sample by Biuret Method

Principle

This is the most commonly used method for the estimation of proteins. Under alkaline conditions copper ions present in Biuret reagent interact with -CO-NH- groups present in adjacent dipeptide and form a purple colored complex. This can be measure spectrophotometrically at 540 nm. The method is fairly specific and there is little interference with other compound containing –CO-NH- groups like urea and other like reducing sugar can interact with cupric ions.

Reaction

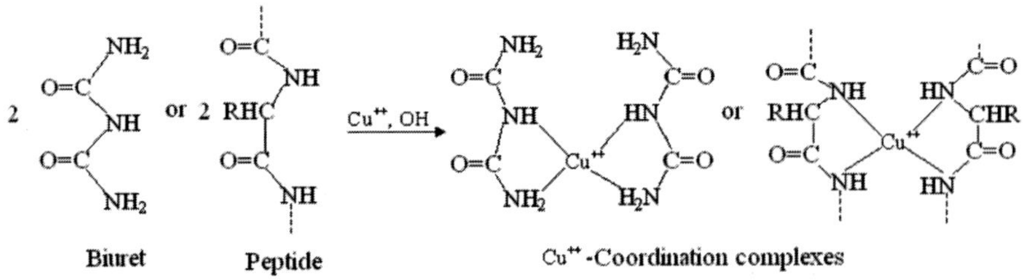

Requirements

Reagents

1. Standard Protein solution (Bovine Serum Albumin (BSA) 5mg/ml): Dissolve 500 mg of BSA and make up the final volume 100 ml with D/W.
2. Biuret reagent: Dissolve 3g of $CuSO_4$ and 9gm sodium potassium tartarte in 500ml of 0.2 M NaOH. Add 5gm of Kl and make up the final volume one liter with 0.2 M NaOH.
3. D/W
4. Test sample

Equipments: Balance, water bath, spectrophotometer

Glassware : Test tubes, concal flasks, glass pipette.

Range of the method: 2-20 mg

Working concentration: 5mg/ml.

Procedure

1. Take clean and dried test tubes. Mark all the tubes as per the protocol table.
2. Take all the reagents along with protein (std) solution with 1, 2, 3 up to 5mg and test solution sequentially as per the protocol table.
3. Make up the final volume 2ml with D/W along with the blank tubes.
4. Add 3ml Biuret reagent in all tubes. Mix all the reagents properly and incubate the mixture as per the protocol table.
5. Measure the color complex spectrophotometrically at 520nm.
6. Draw the standard curve of Concentration of BSA vs Optical Density (O.D.)
7. Put the O.D. of the unknown solution and calculate out the concentration of the protein present in given sample from standard graph and formula (given below) and express your result in mg%.

Observation

Concentration of std = 5mM/ml

Tubes	Aliquote (ml)	Conc (mM)	D/W (ml)	Biuret Reagent		OD at 520 nm
B1	-	-	2.0	3.0	Keep at 37ºC for 15′	
S1	0.2	1.0	1.8	3.0		0.074
S2	0.4	2.0	1.6	3.0		0.113
S3	0.6	3.0	1.4	3.0		0.186
S4	0.8	4.0	1.2	3.0		0.236
S5	1.0	5.0	1.0	3.0		0.291
U1	0.5		1.5	3.0		0.065
U2	1.0		1.0	3.0		0.133

Formula

$$\frac{\text{OD (test)}}{\text{OD (std)}} \times \frac{\text{Conc (std)}}{\text{Aliquot (test)}} \times 100$$

Calculation

1. Directly from graph

Conc of U1 at .065 OD = 1.1

Conc of U2 at .133 OD = 2.3

$\therefore$ U1 = 1.1mg/.5ml = 2.2 mg/ml = 220mg%

U2 = 2.3mg/ml = 230mg%

$$U = \frac{220+230}{2} = 225 \text{ mg\%}$$

2. By slope method

Slope = .08/1.5 = .05

$$U1 = \frac{\text{OD (test)}}{\text{Slope}} = \frac{0.65}{0.05} = 1.3\text{mg}/.5\text{ml} = 260\text{mg}/100\text{ml}$$

$$= 260\text{mg\%}$$

$$U2 = \frac{\text{OD (test)}}{\text{Slope}} = \frac{.133}{.05} = 2.46\text{mg}/1\text{ml} = 246\text{mg\%}$$

$$U = \frac{260+246}{.2} = 253\text{mg\%}$$

3. Using formula

U1 = .065/.291 x 5/0.5 x 100 = 223.3mg%

U2 = .133/.291 x 5/1 x 100 = 228.5mg%

$$U = \frac{223.3+228.5}{.2} = 225.9\text{mg\%}$$

$$\overline{U} = \frac{225+253+225.9}{.2} = 234.6\text{mg\%}$$

Conclusion/Interpretation

The conc. of proteins (BSA) is 234.6 mg%

Concentration

(a) By graph = **225 mg%**

(b) By slope method = **253 mg%**

(c) By formula = **225.9 mg%**

Experiment – 8 : Estimate the Concentration of Protein in given sample by Folin-Lowry's Method

Principle

The aromatics amino acids (tyrosine and tryptophan) present in proteins react with phosphomolydic acid and tungstate present in Folin - Ciocalteau reagent and form a blue colored complex which can be measure spectrophotometrically at 660 nm. This method is 10 times more sensitive than Biuret method.

Requirements

Reagents

1. Standard protein solution (Bovine Serum Albumin (BSA) 250 µg/ml): Prepare stock solution (1mg/ml) and dilute with D/W to make final concentration 250 µg/ml.
2. Folin-Ciocalteau reagent: 2N Folin-Ciocalteau reagent is available commercially. Dilute it to 1:1 proportion for use. Prepare fresh and keep in brown bottle.
3. Reagent A: 2% Sodium carbonate in 0.1N NaOH
4. Reagent B: 0.5% $CuSO_4$ and 1% Sodium potassium tartarate in 100ml D/W.
5. Reagent C: Mix 50ml Reagent A with 1ml Reagent B. Prepare always fresh.
6. D/W
7. Test sample

Equipments: Balance, spectrophotometer.

Glassware : Test tubes, concical flasks, glass pipette etc.

Range of method: 50-250µg.

Working concentration: 250µg/ml.

Procedure

1. Take clean and dried test tubes. Mark all the tubes as per the protocol table.
2. Take all the reagents along with protein (std) solution with 50, 100, 150 up to 250µg and test solution sequentially as per the protocol table.
3. Make up the final volume 4ml with D/W along with the blank tubes.

4. Add 5.5ml alkaline reagent (Reagent C) in all tubes. Mix all the reagents properly and incubate the mixture as per the protocol table.
5. Add 0.2 ml Folin-Ciocalteau reagent in all tubes. Mix well and incubate the mixture as per the protocol table.
6. Measure the color complex spectrophotometrically at 660 nm.
7. Draw the standard curve for Concentration of BSA vs Optical Density (O.D.).
8. Put the O.D. of the unknown solution and calculate out the concentration of the protein present in given sample from standard graph and formula (given below) and express your result in mg%.

Observation

Conc of std: 250µg/ml

Tubes	Aliquot (ml)	Conc (µg)	D/W (ml)	Alkaline Reag (ml)		Folin Reag (ml)		OD at 660 nm
B	-	-	4	5.5	Incubate at room temp for 10′	.2	Incubate at 37^{0}C for 20′	
S1	.2	50	3.8	5.5		.2		.064
S2	.4	100	3.6	5.5		.2		.085
S3	.6	150	3.4	5.5		.2		.110
S4	.8	200	3.2	5.5		.2		.152
S5	1.0	250	3.0	5.5		.2		.174
U1	.5		3.5	5.5		.2		.087
U2	1.0		3.0	5.5		.2		..121

Formula

$$\frac{\text{OD (test)}}{\text{OD (std)}} \times \frac{\text{Conc (std)}}{\text{Aliquot (test)}} \times \frac{100}{1000}$$

Calculation

1. Directly from graph

Conc of U1 at .087 OD = 127

Conc of U2 at .121 OD = 175

$U1 = 127\mu g/.5ml = 254\mu g/ml \quad = 25.4mg\%$

$U2 = 175\mu g/ml \quad = 17.5mg\%$

$$U = \frac{25.4 + 17.5}{2} = \mathbf{21.45mg\%}$$

2. By slope method

$Slope = .055/85 = 6.5 \times 10^{-4}$

$$U1 = \frac{OD\ (test)}{Slope} = .087/6.5 \times 10^{-4} = 133.84\mu g/.5ml$$

$$= 26.7mg\%$$

$$U2 = \frac{OD\ (test)}{Slope} = .121/6.5 \times 10^{-4} = 186.15\mu g/ml$$

$$= 18.6mg\%$$

$U = \mathbf{22.65mg\%}$

3. By using formula

$U1 = .087/.174 \times 250/.5 \times 100/1000 = 25mg\%$

$U2 = .121/.174 \times 250/1.0 \times 100/1000 = 17.4mg\%$

$$U = \frac{25.4 + 17.4}{2} = \mathbf{21.2mg\%}$$

Result

The conc of protein is :

(a) By using method = 21.45 mg%

(b) By slope method = 22.65 mg%

(c) By using formula = 21.2 mg%

Conclusion/Interpretation

By performing folin-howrey's method, we can estimate the concentration of protein in given sample.

Concentration :

(a) By graph = **21.45 mg%**

(b) By slope method = **22.65 mg%**

(c) By using formula = **21.2 mg%**

Experiment – 9 : Estimate the Concentration of Protein in given sample by Bradford's Method

Principle

This mechanism is based on the protein-dye binding method. The binding of Commassie Brilliant Blue (CBB) G-250 to protein in acidic conditions shift the λ_{max} of dye from 465nm to 595nm. Absorption of the blue colored protein-dye complex at 595nm is directly related to concentration of protein present in the sample. The method is more sensitive than the Biuret and Folin-Lowry's method. But, the formed colored complex is unstable with the time interval.

Reaction

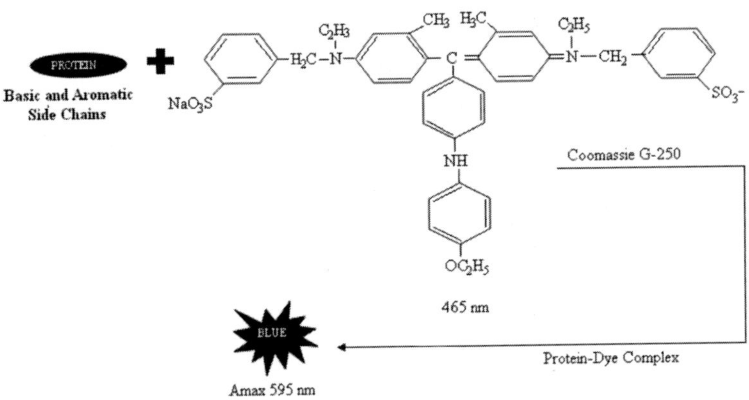

Requirements

Reagents

1. Standard Protein solution (BSA) 20µg/ml: Prepare stock solution (1mg/ml) and dilute with D/W to make concentration 20µg/ml.
2. Bradford reagent: Dissolve 100mg of CBB-G250 in 50ml of ethanol. Cover it tightly and incubate for an hour. Slowly add 100ml of 85% Ortho-Phosphoric acid and make the volume to 1 liter with D/W. Filter the reagent through Whatman filter paper no.1 and Store in brown bottle. Reagent is stable only for 2-3 weeks.
3. 0.2M Phosphate Buffer (pH 7.5): Prepare 0.2M monobasic sodium phosphate and 0.2 M dibasic sodium phosphate and mix until the pH becomes 7.5.

Equipments: Balance, pH meter, Spectrophotometer.

Glassware : Test tubes, conical flasks, glass pipette etc.

Range of the method: 4-20μg

Working concentration: 20μg/ml

Procedure

1. Take clean and dried tubes. Mark all the tubes as per the protocol table.
2. Take all the reagents along with protein (std) solution with 4, 6, 8 upto 20μg and test solution sequentially as per the protocol table.
3. Make up the final volume 2.0 ml with phosphate buffer and add 2ml Bradford reagent in each tube.
4. Mix all the reagents properly and incubate the mixture as per the protocol table.
5. Measure the color complex spectrophotometrically at 595nm.
6. Draw the standard curve for Concentration of BSA vs Optical Density (O.D.).
7. Put the O.D. of the unknown solution and calculate out the concentration of the protein present in given sample from standard graph and formula (given below) and express your result in mg%.

Observation

Conc of std: 50μg/ml

Tubes	Aliquote	Conc (μg)	Phosphate Buffer (ml)	Brod ford Reagent (ml)		OD at 595 nm
B	-	-	2.0	2.0	Incubate for 15′ at room temperature	
S1	.2	10	1.8	2.0		.108
S2	.4	20	1.6	2.0		.172
S3	.6	30	1.4	2.0		.241
S4	.8	40	1.2	2.0		.356
S5	1.0	50	1.0	2.0		.423
U1	.5		1.5	2.0		.174
U2	1.0		1.0	2.0		.297

Formula

$$\frac{\text{OD (test)}}{\text{OD (std)}} \times \frac{\text{Conc (std)}}{\text{Aliquot (test)}} \times \frac{100}{1000}$$

Calculations

1. By direct Method

Conc. of U1 at .174 OD = 22

Conc. of U2 at .297 OD = 34

U1 = 22 µg/.5ml = 44 µg/ml

= 4.4 mg/100ml = 4.4mg%

U2 = 34 µg/ml = 3.4 mg%

$$U = \frac{4.4 + 3.4}{2} = \mathbf{3.9\ mg\%}$$

2. Using formula

U1 = .174/.172 x 20/.5 x 100/1000 = 4.04 mg%

U2 = .297/.172 x 2.0/1.0 x 100/1000 = 3.45 mg%

$$U = \frac{4.4 + 3.45}{2} = \mathbf{3.7\ mg\%}$$

3. By slope method

Slope = dy/dx = .18/21 = 8.5×10^{-3}

$$U1 = \frac{\text{OD (test)}}{\text{Slope}} = .174/8.5\times10^{-3} = 20.4\ \mu g/.5 = 40.8\ \mu g/ml$$

$$= 40.8\ mg\%$$

$$U2 = \frac{\text{OD (test)}}{\text{Slope}} = \frac{.297}{8.5\times10^{-3}} = 34.9\ \mu g/ml = 3.5\ mg\%$$

$$U = \frac{4.08 + 3.5}{2} = \mathbf{3.8\ mg\%}$$

Conclusion/Interpretation

The concentration of protein is :

(a) Bydirect method = **3.9 mg%**

(b) By slope method = **3.7 mg%**

(c) By using formula = **3.8 mg%**

Experiment – 10 : Estimation of Amino Acid by Spectrophotometric Method

Principle

This test is due to a reaction between α- amino group of free amino acid and of ninhydrin i.e. tri-ketohydrin decihydrate. Ninhydrin is a powerful oxidizing reagent in its presence amino acid undergoes oxidative deamination liberating NH_4^+, CO_2 corresponding aldehyde, reduced form of ninhydrin and its reduced products.

Reaction

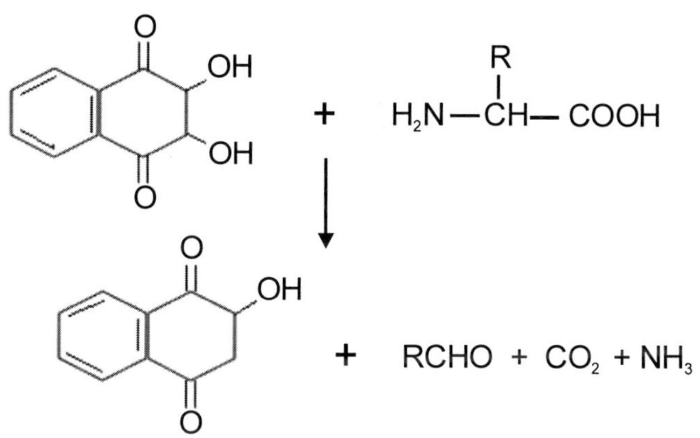

Hydrindentin

Requirements

Reagents

1. Ninhydrin Reagent Solution: 0.35 g ninhydrin in 100 ml ethanol (0.2% sol in acetone)
2. Standard Alanine Solution: 50 µg – L – alanine in 100 ml single distilled water (50 µg/ml of ind. A.A's).

Glass/Plastic wares: test tubes, beakers, measuring cylinder.

Equipments: water bath, spectrophotometer.

Procedure

1. 11 test tubes are taken including one blank.

2. Alanine solution is taken in varying concentration in each and every test tube.
3. Volume is made up to 5 ml with SDW.
4. 1 ml ninhydrin is added to each tube. The tubes are cooked in boiling water bath for 5 minutes and they are cooled at room temperature.
5. O.D. is taken at 570 nm.

Observation

Test Tube	Alanine (Ml)	Single H_2o (Ml)	Ninhydrin (Ml)		O.D. At 570 Nm
Blank	0	5.0	1	INCUBATION IN WATER BATH FOR 5 MIN.	
2	1.0	4.0	1		
3	1.5	3.5	1		
4	2.0	3.0	1		
5	2.5	2.5	1		
6	3.0	2.0	1		
7	3.5	1.5	1		
8	4.0	1.0	1		
9	4.5	0.5	1		
10	5.0	0	1		

Experiment – 11 : Native Disc-Gel Electrophoresis of Proteins

Principle

In native disc gel electrophoresis, the gel consists of different buffer ions. It consists of two types of gels a large pore "Stacking gel" which is polymerized on the top of small pore "resolving gel". The buffers used to prepare these two gels are of different ionic strength & pH. Here, relatively large volume of dilute solution of protein samples can be applied onto the gel and yet good resolution of sample components can be obtained. This is because the proteins in the applied sample get concentrated into extremely narrow zone or get stacked during migration through large poured stacking gel prior to this entry and separation in small pored resolving gel. The resolving gel offers uniform voltage gradient and pH and the proteins are separated according to their intrinsic charge & size.

Requirements

Reagents

1. 40% (w/v) Sucrose
2. 0.01% Bromophenol blue solution
3. 1.5% (w/v) Ammonium per sulphate (APS) Sol to be prepared fresh
4. Acrylamide, Bisacrylamide solution in 30:0.8 ratios. Store in Brown bottle at 0-4^{0}C
5. Stacking gel buffer stock (Tris HCl – pH = 6.8)

   Tris – 6 g, 1 M HCl – 48 ml. Adjust pH to 6.8 and make its final volume to 100 ml with its final voloume to 100 ml with water.
6. Resolving gel buffer stock (Tris HCl – pH = 8.8)

   Tris = 36.3 g, 1 M HCl = 48 ml

   Adjust pH = 8.8 and make its final volume to 100 ml
7. TEMED = (N, N, N′, N′ – Tetramethyl diamine)
8. Reservoir buffer (Tris- Glycine, pH = 8.3)

   Tris = 3g, Glycine = 14.4 g

   Adjust pH to 8.3 & make final volume to 1 ltr. with distilled H_2O.
9. Staining solution

   Commasie Brilliant Blue C -250, -125 g

   Methanol = 200 ml

Glacial Acetic Acid = 35 ml

Make final volume to 500 ml distilled H_2O.

10. Destaining solution:

    Glacial acetic acid = 75 ml

    Methanol = 50 ml

    Mix the components and make final vol. to 1 liter with water.

Glass/Plastic wares: beakers, test tubes, measuring cylinder, glass plates

Equipments: electrophoresis chamber, refrigerator.

Gel Preparation

Reagent	Stacking Gel (2.5%) Ml	Resolving Gel (7.5%) Ml
Acrylamide- Bisacrylamide stock solution	2.50	7.5
Stacking gel buffer	5.00	-
Resolving gel buffer	-	3.75
1% APS	1.50	1.50
Distilled H_2O	11.00	17.25
TEMED	0.015	0.15

Procedure

(a) Sample preparation

To 0.2 ml of solution of the sample preparation containing about 3-4 mg protein 1 ml , add 0.1 ml sucrose and 0.1 ml bromophenol blue solution just prior to loading of the sample in place of sucrose, a drop of glycerol (10%) may be used.

(b) Preparation of gel

1. Thoroughly clean and dry the glass plate and place the whole assembly in an upright position.
2. Mix all the components of resolving gel, except TEMED and APS. Degas the mix for 1 min, using a water pump & then add APS & TEMED in required quantities. Mix gently but quickly.
3. Mix gently & pour the gel solution into the mould in blw the clamped glass plates, taking care to avoid entrapment of any

bubble, overlay the distilled water on the top as gently as possible and leave for 30 min for setting of gel.

4. When the gel has polymerized, remove the water layer & rinse the gel surface with stacking gel.
5. Pour the stacking gel & immediately insert the plastic bomb in the stacking gel.
6. Allow the gel to polymerise for 20 min & then remove the comb.

Electrophoresis of sample

1. Load 10-20 μl of sample in the wells.
2. Switch on the current maintaining it at 10-15 MA. Then increase the current to 30 MA until the bromophenol blue dye reaches the other end.
3. After electophoresis, turn off and disconnected power supply & carefully remove the gel slab in blw glass plates.
4. Place the gel containing staining solution dye for 3-4 hrs or it can be kept for staining overnight and then destain the gel till the clear background of gel is obtained.

Precautions

1. The gel solution should be deaerated.
2. Solution should be stored at 0-4^{0}C.
3. There should be no air bubble in the gel.

Observations

....

....

....

....

Experiment – 12 : Separation of Proteins by SDS-PAGE

Principle

Any charged ion or molecule migrates when placed in an electric field. The rate of migration of a compound depends on its net charge, size and shape & the applied current. This can be given by:

$$V = \frac{Eq}{F}$$

Where, V = velocity of Migration of Molecule

E = Electric field in volt 1 cm

q = The net electric charge on molecule

The movement of a charged molecule in an electric field is often expressed in terms of electrophoretic mobility, which is defined as the velocity per unit the electric field.

$$\mu = \frac{V}{E} = \frac{E}{F} = \frac{q}{E}\left(\mu V = \frac{Eq}{F}\right)$$

For, a molecule with similar conformation varies with size but not with the shape. Thus, electrophoretic mobility (μ) of a molecule is directly proportional to the charge density (Charge/mass).

Molecule with different charge / mass ratio migrates under electric field at different rates and hence gets separate. This is the underlying basic principle of all electrophoretic technique.

Gel Preparation

Stock Solution	Stacking Gel (Ml) 2.5%	Resolving Gel		
		20%	10%	5%
Sol A	2.5	20	10	5
Sol B	-	3.75	3.75	3.75
Sol C	5.0	-	-	-
Sol D	0.2	0.3	0.3	0.3
Sol E	1.0	1.5	1.5	1.5
Sol F	0.015	0.015	0.015	0.015
DDW	11.3	4.45	14.45	19.45

Requirements

Reagents

1. Stock Solution A: 30% w/v acrylamide solution containing 0.8% w/v bisacrylamide (50 g + 0.4 g – 50 ml) in black bottle.
2. Solution B: Reservoir gel buffer

   3 M tris HCl, pH = 8.8 (3.63 g in 4 ml) 4.7 ml HCl

   Total vol. = 10 ml with H_2O
3. Solution C: Stacking gel buffer

   0.5 M Tris HCl (pH = 6.8) (0.69)
4. Solution D: 10% w/v SDS (10 g – 100 ml, 1 g – 10 ml)
5. Solution E: 1.5% w/v APS (1.5 g – 100 ml, 0.05 g – 10 ml)
6. Solution F: TEMED
7. Electrophoretic buffer: 0.025 M Tris HCl, (3 g)

   0.192 M Glycine, (14.4 g)

   & 1% SDS (pH = 8.3) (1 g)
8. Tracking Dye: 0.002% Bromophenol blue in 10% Glycerol / Sucrose
9. Staining Solution: Commasie Brilliant Blue (R-250)

0.1% in water	Methanol	Glacial acetic acid
5	5	2 v/v

   Filter through whatmann filter paper. Store at room temperature.
10. Destaining solution: 7% Acetic acid in water.
11. Sample: Solublising Buffer: 2x pH = 6.8

    125 mm Tris HCl, 4% SDS

    5% bromo phenol blue, 20% sucrose glycerol (pH = 6.8)

Glass/Plastic wares : beakers, test tubes, measuring cylinder, glass plates

Equipments : refrigerator, The components of vertical seat gel electrophoresis apparatus consists of:

(a) Two glass plates, with one of the glass plate being notched & having vertical glass sphere of 1.5 mm in thickness on its side.

(b) Gel containing assembly.

(c) Tygen tubing comb with teeth.

(d) Electrophoretic tank with upper and lower buffer reservoir.

(e) Electrophoretic lead.

(f) Power-pack (0.5 volts)

Procedure

1. Slab gel preparation

(a) Thoroughly clean & dry the glass plate with the help of tygen tubing clamps them & places the whole assembly in an upright position.

(b) Mix various components of resolving gel except SDS. APS & TEMED degas the solution for a min. using a water pump & then add the above remaining component gel.

(c) Mix gently & pore the gel sol into the mould in between the clamped glass plate, taking care to avoid entrapment of any bubble, overlay the distilled water on the top as gently as possible and leave for 30 min. for setting the gel. When the gel has polymerized remove the water layer & rinse the gel surface with stacking gel.

(d) Pour the stacking gel and immediately insert the plastic comb in stacking gel.

(e) Allow the gel to polymerise for 20 min. & then remove the comb.

2. Sample preparation

Mix protein sample (from soaked pulses) with equal volume of sample solubilising buffer. Boil the mixture for 3 mins. in boiling water bath & cool at room temperature.

(Take 1 g of leaf sample & add 2 ml & extration buffer + 6 µl of β-mercaptoethanol). Homogenize sample in motor & pestle centrifuge at 10000 rpm for 5 min & discard pellet & store supernatant at 4^0C.

3. Electrophoresis of sample

(a) Load 10-20 µl samples in the well.

(b) Switch on the current, maintaining it at 10-15 MA, then increase the current to 30 MA until the Bromophenol Blue Dye reaches the other end.

(c) After the electrophoresis is completed twin off and disconnect power supply and carefully remove the gel slab in between glass plates.

(d) Place the gel in a trough containing staining solution dye for 3-4 hrs or it can be kept for staining overnight & then distain the gel till the clear background of gel is obtained.

Observations

....

....

....

....

Experiment – 13 : SDS- Polyacrylamide slab Gel Electrophoresis of Proteins under Reducing Conditions (SDS-PAGE)

Principle

SDS-PAGE is the most widely used method for qualitatively analyzing any protein mixture, monitoring protein purity and to determine their molecular weights. It is based on the separation of proteins according to their size and then locating them by binding to a dye.

SDS or sodium dodecyl sulphate is an anionic detergent that binds strongly to proteins, causing their denaturation. In the presence of excess SDS, about 1.4 g of the detergent binds to each gram of protein, giving the protein a constant negative charge per unit mass. As a result, protein-SDS complex move towards the anode during electrophoresis and owing to molecular sieving properties of the polyacrylamide gel, get separated based on their molecular weights. Since principle of this technique is separated of proteins based on size differences, by running standard proteins of known molecular weights on the same gel as unknown protein, molecular weight of the unknown protein can be determined. Mobility of protein in SDS gel electrophoresis is expressed as relative mobility (R_f) with respect to the tracking dye, bromophenol blue.

R_f= (distance migrated by protein) / (distance migrated by tracking dye)

R_f values of protein marker of known size is used to generate a standard curve by plotting the molecular weights against the R_f value on a semi-log graph. The molecular weights against the R_f value on a semi-log graph. The molecular weight of the unknown protein can then be extrapolated from its R_f value.

The technique consists of three basic steps:

Step I: Preparation of polyacrylamide gels

Crosslinked polyacrylamide gels are formed by co-polymerization of acrylamide monomer and a crosslinking agent N, N'- Methylene bisacrylamide. This reaction is catalyzed by N, N, N', N'-Tetramethylenediamine (TEMED) and initialed by ammonium persulphate (APS). Porosity of the gel is determined by the amount of acrylamide and bis-acrylamide mix used. Lowe percentage gel have larger pore sizes, thereby offering less resistance to passage of larger molecules, while higher percentage gels favour separation of smaller molecules. Most protein separations are carried out using gels ranging from 5-15% acrylamide.

Step II: Electrophoresis

The polyacrylamide gel slab is prepared and fixed to a vertical electrophoresis apparatus. Protein samples are usually denatured by boiling in sample loading buffer containing Tris buffer of pH 6.8, SDS, β-mercaptoethanol (reduces disulphide bonds), sucrose/glycerol (to increase the density), and bromophenol blue (as a tracking dye) and loaded onto wells in the gel.

Resolution of the protein bands is greatly increased by applying the samples onto a short stacking gel top of the separating gel. Differences in pH and composition between these two gels cause the samples to be concentrated into narrow bands by isotachophoresis. As the samples migrate through the separating gel, proteins get resolved depending on their molecular weights. Electrophoresis is stopped when the dye front reaches the bottom of the gel.

Step III: Visualization of proteins

Generally, proteins are colourless and hence cannot be visualized directly. Suitable dyes are used to stain the samples, e.g., Coomasie Brilliant Blue R-250.

This technique is done:

1. To learn preparation of sodium dodecyl sulphate polyacrylamide gel (SDS-PAGE).
2. To determine molecular weight or 'pure' protein by SDS-PAGE.

Requirements

Reagents

Reservoir buffer (Tris-glycine, pH 8.3)

Tris	3.0g
Glycine	14.5g
SDS	1.0g

1. Adjust pH to 8.3 and add water to make final volume to 1 l.
2. SDS (10%, w/v): Dissolve 1 g SDS in 10 ml of distilled water. Store the reagent at room temperature (RT).

3. Sample buffer 2X: The sample preparation buffer which contains a two fold concentration of various components is prepared as given below:

1 M Tris. HCl, pH 6.8	12.5 ml
SDS	4.0 g
b mercaptoethanol	10.0 ml
Glycerol	20.0 ml
1 % Bromophenol blue	4.0 ml

Add water to make final volume to 100 ml.

4. Standard molecular weight marker proteins: The standard molecular weight protein kits, having different molecular weight ranges, are commercially available.

Protein	Molecular weight (kDa)
β-Lactoglobulin	18.4
Trypsinogen	24.0
Carbonic anhydrase	29.0
Pepsin	34.7
Egg albumin	45.0
Bovine serum albumin	66.0

Dissolve 1 mg of each of the above proteins in 1 ml of the sample buffer diluted with water in ration of 1:1 About 20 μl of this mixture should be loaded into one of the wells.

5. Stacking and resolving gels: Stacking and resolving gels for SDS-PAGE are prepared according to the particular given in the following table:

S.No.	Stock solution	Stacking gel (12.5%) ml of the solution	Resolving gel (12.5%)
1	Arcylamide-bis-arcylamide (30:0.8)	2.50	12.50
2	Stacking gel buffer	5.00	-
3	Resolving gel buffer stock solution (Tris-HCl, pH 8.8)	-	3.75
4	10% SDS	0.20	0.30
5	1.5% APS	1.00	1.50
6	Water	11.30	11.95
7	TEMED	0.015	0.015
	Total volume	20.00	30.00

Rocker (optional), measuring cylinder, staining tray, alcohol, distilled water, clamps, gloves, micropipette, tips, water bath.

Glass/Plastic wares: beakers, test tubes, measuring cylinder, glass plates

Equipments

1. Vertical slab gel electrophoresis apparatus. The various items include
   (a) Two glass plates, with one of the glass plates being notched and having vertical glass spacers of 1.5 mm in thickness on its sides
   (b) Gel casting assembly
   (c) Comb with teeth
   (d) Electrophoretic tank with upper and lower buffer reservoirs
   (e) Electrophoretic leads
2. Power pack (0-500 volts).

Procedure

(a) Sample preparation

Mix the protein sample with equal volume of sample buffer. Boil the mixture for 3 min in a boiling water bath and cool to room temperature. If protein sample is very dilute, precipitate the proteins with 10% TCA (incubate in ice for 30 min), centrifuge at 10,000 x g for 5 min. Wash the precipitate with ethanol: ether (1:1) to remove TCA. Dissolve the precipitate in the sample buffer diluted with water in 1:1 ratio, boil for 3 min and cool.

(b) Preparation of slab gels

1. Thoroughly clean, dry the glass plates and assemble them in gel casting assembly. Clamp them and place the whole assembly in an upright position.
2. Mix various components of resolving gel as indicated earlier except for SDS, APS and TEMED. Degas the solution for 1 min using a water pump and then add the above remaining components of the gel.

3. Mix gently and pour the gel solution into the mould in between the clamped glass plates taking care to avoid entrapment of any air bubbles. Overlay distilled water on the top as gently as possible and leave for 30 min for setting of the gel.
4. When the gel has polymerized, remove the water layer and rinse the gel surface with stacking gel buffer.
5. Mix the stacking gel components in the same way as described above for the resolving gel.
6. Pour the stacking gel and immediately insert the supplied plastic comb in the stacking gel. Care should be taken that no air bubbles are entrapped. Allow the gel to polymerize for about 20 min.
7. After the stacking gel has polymerized, remove the comb without distorting or damaging the shapes of the wells. Clean the wells by flushing with electrode buffer using a syringe.
8. Remove the tygon tubing around the two plates and install the gel plate assembly into the electorphoretic apparatus. Pour reservoir buffer in the lower and upper chambers. Remove and trapped bubbles at the bottom of the gel.

(c) Electrophoresis of sample

1. Load 10-20 µl sample (100-200 µg protein) in the sample wells. Also load molecular weight marker proteins in one or two of the wells.
2. Switch 'ON' the current maintaining it at 10-15 mA for initial 10-15 min until the samples have travelled through the stacking gel. Then increase the current to 30 m A until the bromophenol blue dye reaches near the bottom of the gel slab. This may require 3-4 h.
3. After the electrophoresis is complete, turn 'OFF' and disconnect the power supply and carefully remove the gel slab from in between the glass plates.
4. Place the gel in a trough containing solution for 3-4 or it can be kept for staining overnight. Distain with distaining solution till a clear background of the gel is obtained.
5. Record the distance travelled by the dye and the various protein in bands.

6. Plot a graph between $\log_{10}$ molecular weight versus R_m values of standard molecular weight markers. A straight line would be obtained. From the R_m values of the sample polypeptides determine their molecular weights using the above standard calibration curve. Use a ruler to measure the distance migrated by each band of the protein marker from the start of separating gel.
7. Similarly, measure the migration distance of protein sample and the tracking dye from the start of separating gel.
8. Calculate the relative mobilities of the proteins (marker and sample) as follows:
9. R_m = (distance migrated by protein) / (distance migrated by tracking dye)
10. Record results in the table and draw standard curve for estimation of molecular weight of protein. R_m value (y-axis) vs. Molecular weight of protein (kd) (x-axis).

Observations

....

....

....

....

Experiment – 14 : Estimation of total Nitrogen by Micro – Kjeldahl Method

Principle

This method is based on the conversion of the protein and other nitrogenous compounds into ammonium sulfate by boiling with conc. H_2SO_4, subsequent decomposition of the ammonium sulfate by means of a fixed alkali (NaOH); and collection of the liberated ammonia in an acid of known strength. Finally this dissolved ammonia solution is titrated with an acid of known strength and the nitrogen content of the sample calculated.

The reactions involved can be illustrated by reference to glycine:

$$NH_2CH_2COOH \xrightarrow[\text{Heat}]{H_2SO_4} (NH_4)_2SO_4$$

$$(NH_4)_2SO_4 + 2NaOH \longrightarrow Na_2SO_4 + 2NH_3 + H_2O$$

The ammonia can be absorbed in saturated aqueous boric acid solution (this contain about 4% of boric acid).

$$NH_3 + H_3BO_3 \text{ --------- } NH_4^+ + H_2BO_3^-$$

The ammonium borate formed can be titrated directly as an alkali with HCl or H_2SO_4, using screened methyl red as an indicator.

$$H_2BO_3^- + H^+ \text{ --------- } H_3BO_3$$

Requirements

Reagents

1. Conc. H_2SO_4.
2. Catalytic mixture - Mix 3 parts of $CuSO_4$ with one part of selenium metal with anhydrous sodium sulfate (5 g of sod. sulfate and 1 g of selenium metal).
3. 4% Boric Acid: 4 g of boric acid in 100 ml distilled water. Solution should be boiled for a while and cooled to drive away dissolved CO_2.
4. Indicator: Dissolve 1 g of bromocresol green and 0.2 g of methyl - red in 100 ml of 95% ethyl alcohol. Adjust this solution to mid - colour (dark, green, pH 4.2) with 0.1 N NaOH. The bluish green colour of the indicator charges sharply to gray at the end point. Alternatively, screened methyl red can also be used as indicator. It is prepared as follows:

Dissolve separately 0.25 g of methyl red and 0.186 g of methylene blue in 100 ml each of ethanol, mix equal volumes of these two solutions and use few drops as indicator. The first appearance of violet colour is taken as the end point.

5. N/14 H_2SO_4: Mix .99 ml of conc. H_2SO_4 with 1 liter of distilled water.
6. 40% Sodium Hydroxide solution.

Glass/Plastic wares: beakers, test tubes, measuring cylinder,

Equipments: micro-Kjeldahl flask, desiccators, dropper.

Procedure

A. **Digestion:** Accurately weigh 100 mg powdered plant material and transfer carefully into a micro – Kjeldahl flask. Add 1 g catalyst to it. Pour 3 ml conc. sulfuric acid into the flask and boil the mixture until it becomes light green or colourless and then for about one hour longer. Use a low flame initially. Simultaneously run a blank without the sample.

B. **Distillation:** Transfer the clear colourless digest into a distillation flask with the help of distilled water. Take 25 ml of 4% boric acid into a beaker add 2 – 3 drops of indicator into it. Arrange the beaker such that the end point of the delivery tube of the distillation assembly is completely dipped into the boric acid solution. Continue the distillation of the distillation flask.

   As the distillation continues, the liberated ammonia collected into the boric acid solution, the colour of letter changes. Continue the distillation for one hour so that all ammonical nitrogen present in the distillation flask has been liberated and trapped into boric acid solution.

C. **Titration:** Titrate the partly neutralized boric acid solution with N/14 H_2SO_4. Calculate the N content of the plant tissue as follows:

$$\% \text{ Nitrogen} = \frac{100 \times (V_1 - V_2)}{W}$$

V_1 = Volume of sulfuric acid consumed in sample

V_2 = Volume of sulfuric acid consumed in flask

W = Weight of samples taken

***Note :** Each ml of H_2SO_4 is equivalent to 1 mg nitrogen in the form of ammonia.*

Standardization of Sulfuric Acid: Take approximately 10 – 20 g sodium carbonate and heat it upto 260 – 270^{0}C for half an hour to remove its moisture. Allow it to cool down in a desiccator before use.

Weight 3.78 g of dried sodium carbonate rapidly. It is important to remember that anhydrous sodium carbonate is hygroscopic and exhibits a tendency to pass into the monohydrate. Dissolve the weighed sodium carbonate in one litre distilled water. Take 25 ml of this solution into a conical flask and add to it few drops of methyl orange indigo carmine indicator.

Rinse a clean burette with N/14 sulphuric acid and fill it up to zero mark. Titrate the sodium carbonate solution by slowly adding the acid dropwise until the colour of the indicator becomes neutral grey. This marks the end point of the titration. Record the burette reading in your note – book. Repeat the titration till matching data are available.

The normality of sulphuric acid solution is calculated according the equation:

$$N_1V_1 = N_2V_2$$

Where:

N_1 = Normality (0.071N) of sodium carbonate solution

V_1 = Volume of sodium carbonate solution (25 ml)

N_2 = Normality of sulfuric acid solution

V_2 = Volume of sulphuric acid solution used in titration (ml)

***Note** :Methyl orange – indigo carmine indicator should be prepared by dissolving 1 g of methyl orange and 2.5 g of purified indigo carmine in 1 ml of distilled water, and filtering the solution. The colour change on passing through alkaline to acid solution is form green to magenta with a neutral grey colour at pH of about 4.0.*

Observations

....

....

....

....

Additional Reading

Barrett, G.C. (editor) (1985). Chemistry and Biochemistry of the Amino Acids. Chapman and Hall, New York.

Boulter, D. (1970). Protein synthesis in plants. Ann. Rev. Plant Physiol. 21: 1.

Bradford, M.M (1976). Anal. Biochem. 72 : 248-256.

Chial, H.J., Congdon, R.W., and Splittgerber, A.G. (1995). J. Chem. Ed. 72 : 76-79.

Cutler, P. (2004). Protein Purfication Protocols. Totowa, NJ: Humana Press. (Detailed theory and practical procedures for a range of purification techniques).

Delauney, A.J. Verma D.P. (1993). Proline biosynthesis and osmoregulation in plants. Plant J. 4 : 215-223.

Greenstein, J.P. and Winitz, M(1961). Chemistry of Amino Acid vol. III John Wiley and Sons. Inc. NY, USA.

Meister, A.(1965). Biochemistry of Amino Acid Vol 1, 2nd ed. Academic Press Inc. NY, USA.

Miflin B.J. ed (1980). The Biochemistry of plants vol 5, Amino Acids and Derivatives. Acedemic Press, New York.

Simpson, R.J., Adams, P.D. and Golemis, E.A. (2008). Basic Methods in Protein Purification and Analysis: A Laboratory Manual. New York: CSH Press. (A comprehensive collection of protein purification methods).

Singh, B.K. Shaner, D.L. (1995). Biosynthesis of branched chain amino acids from test tube to field plant July, 7(7) : 935-944.

Thompson, J.D. (2008). Functional Proteomics. New York: Humana Press. (Comprehensive coverage of functional proteomics including protein analysis and mass spectrometry.)

U.K. Laemmli (1970) Cleavage of structural proteins during the assembly of the head of bacteriophage T_4. Nature. 227:680.

❑❑❑

Unit -4

Lipids

Lipids are a hetrogenous group of substances, widely distributed throughout the plant and animal kingdom. In plants they are present in the seeds nuts and fruits, while in the animals they are stored in adipose tissues, bone marrow and nervous tissues. Lipids are relatively insoluble in water and readily soluble in organic solvents such as ether, chloroform, carbon disulfide, benzene, hot alcohol etc. The term lipid was used by the German biochemist Bloor in 1943 for a major class of tissue components and foodstuffs.

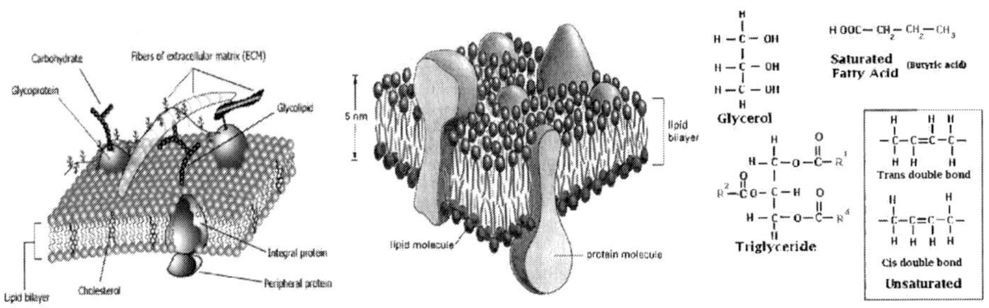

Fig. 4.1 : Lipids molecules

> • *"Whenever science makes a discovery, the devil grabs it while the angels are debating the best way to use it."*
>
> *Alan Valentine*

Experiment – 1 : Extraction and Gravimetric Estimation of lipid from Ground Nut by Soxhlet's Method

Principle

Lipids are soluble in some organic solvents. This property of specific solubility in non-polar solvents such as hexane, benzene, chloroform is utilized for extracting lipids from various tissues. In biological materials the lipids are generally bound to proteins and they are therefore extracted either with a mixture of ethanol and diethyl ether or a mixture of ethanol and methanol, which helps in breaking the bonds between the lipids and proteins.

Requirements

Reagents

1. Ground nut powder: Crush the ground nut homogenously to make the powders (5gm)
2. Non polar solvents: Hexane, Chloroform, Methanol and benzene
3. Thimble Preparation: Pack 5gm of fine crushed ground nut powder in filter paper and tight it with the strings. Put the thimble in extractor of Soxhlet apparatus.

Glass/Plastic wares: Beakers, test tubes

Equipments: Balance, Mortar and Pestle, Soxhlet apparatus

Procedure

1. Perform the soxhlet extraction till all the oil has been extracted (Color of the RBF solvent turn to yellowish black or greenish black).
2. Distill the solvent by removing the thimble and repeat the round of soxhlet extraction.

Calculation

A = Weight of RBF (Preweight)

B = Weight of Dry RBF + Oil contain

So weight of sample B-A = ____________.

$$\% \text{ of oil in Ground nut} = \frac{100 \times \text{Weight of substance}}{\text{g of substance in thimble}}$$

Observations

....

....

....

....

Experiment – 2 : Estimate the Lipid Present in given sample by Sulfo-Phosphovanilin Method

Principle

Lipid reacts with vanillin in presence of concentrated H_2SO_4 and phosphoric acid to form pink colored complex which can be measure spectrophotometrically at 456 nm.

Requirements

Reagents

1. Standard lipid solution (100µg/ml): Dissolve 1 gm ground nut oil in 100 ml of chloroform.
2. Color reagent: Dissolve 2.1gm Vanillin in 350 ml D/W and slowly add 600ml O-phosphoric acid with continous stirring. Make the final volume 1 liter with D/W.
3. Concentrated H_2SO_4.
4. Unknown solution: Prepare homogenous mixture of 5gm fine crushed ground nut powder with 20ml of hexane, crush it well and add hexane if needed. Filter the mixture (till the clear solution is obtained) and make up the final volume to 100 ml with hexane. Use this solution for the estimation of lipid.

Equipments: Balance, Mortar and Pestle, water bath, Spectrophotometer

Glass/Plastic wares: Test tubes, conical flasks, glass pipette.

Range of the method: 20-100µg.

Working concentration: 100µg/ml.

Procedure

1. Take clean and dried test tubes. Mark all the tubes as per the protocol table.
2. Take all the reagents along with lipid (std) solution with 20, 40, 60, up to 100µg and test solution sequentially as per the protocol table.
3. Incubate all the tubes in BWB for the complete evaporation of solvent.
4. Carefully add 2ml concentrated H_2SO_4 with the help of burette in all the tubes mix well and again incubate in BWB as per the protocol table.

5. After cooling measure the color complex spectrophotometrically at 456nm.
6. Draw the standard curve for Concentration of lipid vs Optical Density (O.D.)
7. Put the O.D. of the unknown solution and calculate out the concentration of the lipid present in given sample from standard graph and formula (given below) and express your result in mg%.

Formula

$$\frac{OD(test)}{OD(std)} \times \frac{Conc(std)}{Aliquot(test)} \times \frac{100}{1000}$$

Observations

....

....

....

....

Experiment – 3 : Determination of Acid-Value of Fats and Oils

Principle

The amount of free fatty acid can be determined volumetrically by titrating the sample with KOH. The acidity of fat is expressed as its acid values or number which is defined as milligram of KOH required to neutralize the free fatty acid present in one gram of fat. The amount of free fatty acid present for acid value of fat is useful because it gives an indication about the time and extent of deterioration.

Requirements

Reagents : Fat sample KOH, indicator phenolphthalein, oxalic acid, fat solvent is made up of equimelar mixture of 95% alcohol and ether (1:1).

Glass/Plastic wares: Titration flask, beaker, measuring cylinder, microtips

Equipments: micropipette, weighing balance

Procedure

I. Titration of 0.1 N KOH with 0.1 N oxalic acid

S.no.	Initial Reading (ml)	Final Reading (ml)	Vol. of KOH Used (ml)	Mean
1.	0.0	22 ml	22 ml	
2.	22	44 ml	22 ml	22 ml
3.	44	66 ml	24 ml	

1. The Burette is set with KOH solution.
2. 10 ml 0.1 N oxalic acid is titrated along with 1-2 drops of phenolphthalein.

II. Titration of KOH with fat solution

S.no.	Initial Reading (ml)	Final Reading (ml)	Vol. of KOH Used (ml)	Mean
1.	0.0	0.1	0.1	
2.	0.1	0.3	0.2	0.1 ml
3.	0.3	0.4	0.1	

1. 10 ml of blank solvent is taken in titration flask.
2. 1-2 drops of phenolphthalein is then added to this mixture.
3. 10 ml of solution is taken titrated with 0.1 N KOH.

III. Titration of KOH with fat solution

S.no.	Initial Reading (ml)	Final Reading (ml)	Vol. of KOH Used (ml)	Mean
1.	0.0	0.3	0.3	
2.	0.3	0.6	0.3	0.3 ml
3.	0.6	0.9	0.3	

1. 10 gm of fat sample is mixed with 30 ml of solvent in a flask
2. 1-2 drops of phenolphthalein is then added to this mixture
3. 10 ml of solution is taken out from the above mixture and titrated against 0.1N KOH.
4. Acid value is calculated then.

Calculations

10 ml of 10% fat solution = 1 gm of fat

Volume of KOH used (titration of oxalic acid)

Volume of oxalic acid used = 10 ml

$V_1S_1 = V_2S_{2-}$

(oxalic acid) (KOH)

$10 \times 0.1 = 22 \times S_2$

$S_2 = 1/22$

Normality of KOH = Strength (S_2) x Eq. wt.

$= 1/22 \times 56 \quad = 2.54$ N

0.896 KOH Sol. used for blank (γ) = 0.1 ml

0.896 KOH Sol. used for blank + fat (α) = 0.3 ml

Exact, KOH Sol. used for Titration = ($\alpha - \gamma$) ml

$= (0.3 - 0.1)$

$= 0.2$ ml

1000 ml of solution contains $= \frac{56}{22}$ gm KOH

$$(\alpha - \gamma) \text{ ml contains} = \frac{56}{22} \times \frac{0.2}{1000}$$

$$= \frac{56}{22} \times \frac{0.2}{1000}$$

$$= 0.0005 \text{ gm KOH}$$

Observations

....

....

....

....

Experiment – 4 : Determination of Saponification Value of Fat and Oil

Principle

Hydrolysis of fat with an alkali results in the formation of salts of fatty acids and glycerol. This process is called saponification from the amount of KOH utilized during hydrolysis, the saponification value of given fat sample can be calculated. The saponification value can be defined as milligram (mg) of KOH required to saponify one gram of the given fat.

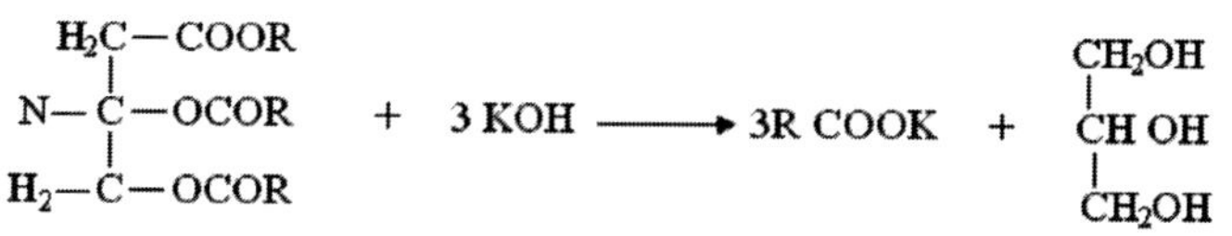

Triacylglycerol **Potassium salt of fatty acid** **Glycerol**

From the above reaction it may be noted that 3 molecules of KOH are consumed for saponification of each molecular of triacylglycerol irrespective of chain length of fatty acid. Evidently each gram of triacylglycerol with contain larger number of molecules of triacylglycerol with thus, require much more KOH. The saponification value is thus, an indicator of average molecular weight of the fatty acid in an acylglycerol.

The procedure involving reflection of known amount of fat with a fixed but an excess of alcoholic KOH, the amount of KOH remaining after hydrolysis is determined by black titrating with the standardized 0.5N HCl and the amount of KOH utilized for saponification can be calculated.

Requirements

Reagents

Fat sample : 0.5 N HCl, 0.5 N alcoholic KOH

Fat solvent : 95% Ethanol and Ether in 1:1 ratio

Glass/Plastic wares : Titration flask, beaker, measuring cylinder, microtips

Equipments : Micropipette, weighing balance

Procedure

I. Titration of HCl with fat solution

S.no.	Initial Reading (ml)	Final Reading (ml)	Vol. of KOH Used (ml)	Mean
1.	0	21.6	21.6	21.6

1. Take 1 gm of fat in 250 ml conical flask and add 10 ml of fat solvent.
2. Then, 25 ml of 0.5 N alcoholic KOH is added to the flask.
3. The flask is kept in water bath.
4. After boiling, the flask is cooled at room temperature and 1-2 drops of phenolphthalein is added.
5. The solution is then titrated against 0.5 N HCl.

II. Titration of HCl with blank solution:

S.no.	Initial Reading (ml)	Final Reading (ml)	Vol. of KOH Used (ml)	Mean
1.	0	17.8	17.8	17.8

1. 10 ml of fat solvent is taken in a flask.
2. 25 ml of alcoholic KOH is added to it and the flask is kept in water bath for 30 min.
3. After cooling the flask, the solution is titrated against 0.5 N HCl.

Calculations

Volume of HCl used for fat solution = 'X' = 21.6 ml

Volume of HCl used for fat Blank solution = 'Y' = 17.8 ml

Exact volume of HCl used only for fat sample

$$= \underset{(HCl)}{N_1V_1} = \underset{(KOH)}{N_2V_2}$$

$$= 0.5 \times 3.8 = 0.5 \times V_2$$

$$V_2 = 3.8$$

Strength of KOH = Normality x Eq. wt.

= 0.5 x 56

= 28.0 gm/ml

100 ml of KOH solution contains = 28000 mg KOH

$$3.8 \text{ ml of KOH solution contains} = \frac{28000}{1000} \times 3.8 \text{ gm of KOH}$$

= 106.4 gm

Observations

....

....

....

....

Experiment – 5 : Estimation of Free Fatty Acids

Principle

The free fatty acid in oil is estimated by titrating it against KOH in the presence of phenolphthalein indicator. The acid number is defined as the mg KOH required neutralizing the free fatty acids present in 1 g of sample. However, the free fatty acid content is expressed as oleic acid equivalents.

Requirements

Reagents :

- 1% phenolphthalein in 95% ethanol.
- 0.1 N potassium hydroxide.
- Neutral solvent: Mix 25 ml ether, 25 ml 95% alcohol and 1 ml of 1% phenolphthalein solution and neutralize with N/10 alkali.

Glass/Plastic wares: Titration flask, beaker, measuring cylinder, microtips

Equipments : micropipette, weighing balance

Procedure

1. Dissolve 1-10 g of oil or melted fat in 50 ml of the neutral solvent in a 250 ml conical flask.
2. Add a few drops of phenolphthalein.
3. Titrate the contents against 0.1 N potassium hydroxide.
4. Shake constantly until a pink colour which persists for fifteen seconds is obtained.

Calculation

$$\text{Acid value(mg KOH / g)} = \frac{\text{Titre value} \times \text{Normality of KOH} \times 56.1}{\text{Weight of the sample (g)}}$$

The free fatty acid is calculated as oleic acid using the equation 1 ml N/10 KOH = 0.028 g oleic acid.

Note : To find out the exact strength of KOH, prepare 0.1 N oxalic acid solution (630 mg in 100 ml water) and titrate against KOH with phenolphthalein as indicator. Calculate the strength of KOH by the formula $V_1N_1 = V_2N_2$.

Observations

....

....

....

....

Experiment – 6 : Determination of Iodine Value of Oil

Principle

The oils contain both saturated and unsaturated fatty acids. Iodine gets incorporated into the fatty acid chain wherever the double bonds exist. Hence, the measure of iodine absorbed by oil, gives the degree of unsaturation. Iodine value/number is defined as the 'g' of iodine absorbed by 100 g of the oil.

Requirements

Reagents

- Hanus Iodine Solution: Weigh 13.6 g of iodine and dissolve in 825 ml glacial acetic acid by heating, and cool. Titrate 25 ml of this solution against 0.1 N sodium thiosulphate. Measure another portion of 200 ml of glacial acetic acid and add 3 ml of bromine to it. To 5 ml of this solution and 10 ml of 15% potassium iodide solution and titrate against 0.1 N sodium thiosulphate. Calculate the value of bromine solution, to double halogen content of the remaining 800 ml of the above iodine solution as follows:

  X = B/C, where X = ml of bromine solution required to double the halogen content, B = 800 x thiosulphate equivalent of 1 ml of bromine solution.
- 15% Potassium Iodide Solution.
- 0.1% Sodium Thiosulphate.
- 1% Starch.

Glass/Plastic wares: Titration flask, beaker, measuring cylinder, microtips

Equipments: micropipette, weighing balance

Procedure

1. Weigh 0.5 or 0.25 g of oil into an iodine flask and dissolve in 10 ml of chloroform.
2. Add 25 ml of Hanus iodine solution using a pipette, draining it in a definite time. Mix well and allow to stand in dark for exactly 30 min with occasional shaking.
3. Add 10 ml of 15% KI, shake thoroughly and add 100 ml of freshly boiled and cooled water, washing down any free iodine on the stopper.

4. Titrate against 0.1 N sodium thiosulphate until yellow solution turns almost colourless.
5. Add a few drops of starch as indicator and titrate until the blue colour completely disappears.
6. Towards the end of titration, stopper the flask and shake vigorously so that any iodine remaining in solution in $CHCl_3$ is taken up by potassium iodide solution.
7. Run a blank without the sample.

Calculation

The quantity of thiosulphate required for blank minus the quantity required for sample gives thiosulphate equivalent of iodine absorbed by the fat or oil taken for analysis.

$$\text{Iodine number} = \frac{(B-S) \times N \times 12.69}{\text{g sample}}$$

where B = ml thiosulphate for blank.

S = ml thiosulphate for sample.

N = normality of thiosulphate solution.

Amount of fat/oil taken should be adjusted such that the excess iodine in the added 25 ml of Hanus iodine solution has about 60% of excess iodine of the amount added, i.e., if (B - S) is greater than B/2, repeat with smaller amount of sample.

Observations

....

....

....

....

Experiment – 7 : Determination of Peroxide Value of Oil

Principle

Peroxide value is a measure of the peroxide contained in the oil. The peroxides present are determined by titration against thiosulphate in the presence of KI. Starch is used as indicator.

Requirements

Reagents

- Solvent mixture: mix two volumes of glacial acetic acid with one volume of chloroform.
- 5% Potassium Iodide Solution.
- 1% Starch Solution.
- N/500 Sodium Thiosulphate Solution. Prepare N/10 solution and dilute to N/500 on the day of use.

Glass/Plastic wares: Titration flask, beaker, measuring cylinder, microtips

Equipments: micropipette, weighing balance

Procedure

1. Weigh 1 g of oil or fat into a clean dry boiling tube and add 1 g of powdered potassium iodide and 20 ml of solvent mixture.
2. Place the tube in boiling water so that the liquid boils within 30 seconds and allow to boil vigorously for not more than 30 seconds.
3. Transfer the contents quickly to a conical flask containing 20 ml of 5% potassium iodide solution.
4. Wash the tube twice with 25 ml water each time and collect into the conical flask.
5. Titrate against N/500 sodium thiosulphate solution until yellow colour is almost disappeared.
6. Add 0.5 ml of starch, shake vigorously and titrate carefully till the blue colour just disappears.
7. A blank should also be set at the same time.

Calculation

Peroxide value (milliequivalent peroxide/kg sample) ,

$$\text{Iodine number} = \frac{S \times N \times 100}{\text{g sample}}$$

where S = ml $Na_2S_2O_3$ (Test-Blank) and N = normality of $Na_2S_2O_3$.

Observations

....

....

....

....

Experiment – 8 : Estimation of Volatile Acids

Principle

Straight distillation or steam distillation after acidifying the sample with 1 + 1 H_2SO_4 results in the evaporation of volatile acids which on condensation yield liquefied volatile acids. These acids could be titrated against 0.1 N alkali with phenolphthalein indicator to find out the amount of volatile fatty acids.

The steam distillation method is more tedious, requiring at least 4 h to complete, but results in the recovery of 92% to 98% of the volatile acids from sewage sludge. The steam distillation is conducted on a $MgSO_4$ – saturated, acidified, sludge-free liquor produced by chemical treatment with ferric chloride and filter – aid followed by vacuum filtration. The effects of varying concentrations of dissolved solids and mineral acids during distillation are avoided.

Requirements

Reagents

- Dilute sulphuric acid 1 + 1: Mix one volume of conc. H_2SO_4 with one volume of water.
- Ferric chloride solution: Dissolve 82.5 g $FeCl_3.6H_2O$ in one liter distilled water.
- Diatomaceous: Silica Filter Aid
- Magnesium sulphate ($MgSO_4.7H_2O$)
- Standard sodium hydroxide, titrant 0.1 N: Dissolve 4 g sodium hydroxide in 1 L of distilled water and standardize against standard acid.
- Phenolphthalein indicator: 0.1% in 95% alcohol

Glass/Plastic wares: Titration flask, beaker, measuring cylinder, microtips

Equipments: micropipette, weighing balance

Procedure

1. Adjust the sample containing volatile fatty acids (sludge and water) 200 – 1000 ml to pH 3.5 with (1+1) H_2SO_4.
2. Add 6 ml of $FeCl_3$ solution/L sample (equivalent to 500 mg/L).
3. Add 50 g filter aid/L, and mix well.

4. Filter by suction, using a Buchner funnel containing a filter paper freshly coated with thin layer of filter - aid.
5. Wash the residue thoroughly three or four times with water, and adjust the filtrate pH 11 with NaOH solution.
6. Concentrate by evaporation to 150 ml and cool in a refrigerator.
7. Adjust the cooled filtrate pH 4 with dilute H_2SO_4 and add it quantitatively and quickly to the distilling flask.
8. Add magnesium sulphate to a slight excess of saturation.
9. Apply heat with a small flame to the flask until rapid evolution of volatile acids commences. This will prevent excessive increase in the volume of the mixture.
10. Steam distills slowly so that about 200 ml of distillate is collected in 25 min.
11. Increase the rate of distillation and continue until a total of 600 ml is collected.
12. Titrate the distillate against 0.1 N NaOH using phenolphthalein indicator.

Calculation

$$\text{Mg/L volatile acids as acetic acid (molecular weight of acetic acid is 60)} = \frac{\text{ml 0.1 N NaOH} \times 6{,}000}{\text{ml sample}}$$

Observations

....

....

....

....

Experiment – 9 : Identification and Quantification of Fatty Acids

Principle

Fatty acids are made volatile by converting them into methyl esters. The conversion of fatty acids in methyl esters is carried out directly by trans esterification i.e., from glycerol ester to methyl ester. Esterification can also be done after saponification. The esters are identified and quantified by injecting into GLC and comparing with a set of standard esters.

Requirements

Reagents

- 10% BCl_3 or 14% BF_3 in methanol
- Saturated NaCl solution
- Na_2SO_4 (anhydrous)
- Hexane
- Nitrogen Gas
- GLC

Column	-	Pretested 10% silar 10°C on gas chrom Q 100 to 200 mesh (6′ x ¼″ – internal diameter 4 mm).
Detector	-	FID
Injector temp.	-	280°C
Detector temp.	-	250°C
Carrier gas	-	Nitrogen at 50 ml/min
Column/oven temp.	-	165°C

Glass/Plastic wares: Titration flask, beaker, measuring cylinder, microtips

Equipments: micropipette, weighing balance

Procedure

1. To 150 – 300 mg of oil, in a culture tube add 3 ml of 10% BCl_3 or 14% BF_3.
2. Add boiling chips, cover the tube with aluminium foil and heat at 83°C for 6 min.

3. Transfer the contents to a 30 ml separating funnel and also collect the washings of the tube into the separating funnel. Washing is done four tunes with 1 ml portions of hexane.
4. Shake and allow to separate.
5. Add 4 ml saturated NaCl solution.
6. Shake and collect hexane layer over anhydrous Na_2SO_4.
7. Rinse funnel with 2 ml hexane, collect the hexane layer and combine the hexane extracts.
8. Filter the hexane extract through filter paper (Whatmen No. 4).
9. Reduce the volume of the filtrate to 2 -3 ml by drying with a slow stream of nitrogen.
10. Inject an aliquot to pre - conditioned GLC.
11. Inject standard methyl esters separately and calculate the retention time for individual esters.

Calculation

Identify peaks by relative position on the chart. Esters appear in order of increasing number of carbon atoms and of increasing unsaturation for the same number of carbon atoms. Calculate the peak area of each acid in the sample and compare with the standards. From the quantity injected and the attention, the amount of fatty acid content can be calculated. The latest computerized GLC models give concentration directly.

Observations

....

....

....

....

Experiment – 10 : Steam Distillation and Isolation of Essential Oils

Principle

Steam is used to volatilize the oil from the source material, and the oil vapour is collected along with water vapour. The oil, after condensation, remains immiscible with water, recovered and analyzed.

Requirements

Reagents

- Plant (Example, jasmine flowers/eucalyptus leaves/ aswagandha) tissue (fresh).
- Sodium Sulphate (anhydrous), Ice, Water

Glass/Plastic wares: Titration flask, beaker, measuring cylinder, microtips, Distillation unit (glass), Conical/Erlenmeyer flask

Equipments : micropipette, weighing balance, Heating element.

Procedure

1. Assemble the distillation unit over heating facility and fill with distilled water.
2. Boil the distilled water in the round bottom flask and allow the steam to pass through the glass tube to the bottle containing the sample (The steam outlet opens at the bottom of the flask sufficiently filled with sample material).
3. The steam boils the volatile oils present in the material and is carried along with the steam.
4. The volatile oil vapour pass through the (Liebig) condensing tube (surrounded by circulating water) and get collected with water in a flask surrounded by ice.
5. The volatile oil being immiscible with water and light weight floats on top.
6. Continue the distillation for 15 – 45 min depending upon the nature and quantity of source material.
7. Remove the collection flask, transfer the content to a suitable separating funnel and allow for 15 min partitioning the immiscible volatile oil from water.

8. Save the oil fraction in an Erlenmeyer flask, add a few g of sodium sulphate (anhydrous) and stand for 30 min to remove any moisture.
9. Measure or weigh the volatile oil fraction and express the % oil content per g tissue.
10. Depending upon the chemical nature of the distillate, it can be subjected to further suitable qualitative/quantitative analysis.

Notes :

1. *All the containers used should be tightly closed with suitable one and two-hole rubber/pith corks.*
2. *Pressure builds up in the steam generating and sample flasks considerably during distillation.*
3. *Set up the distillation in a fume hood or under negative pressure to avoid contamination of lab environment due to, if any, escaping volatile vapours.*

Observations

....

....

....

....

Additional Reading

Asselineau, J. and Lederer, E. (1961). Chemistry of lipids. Ann. Rev. Biochem. 30: 71-92.

Bloor, W.R. (1943). Biochemistry of the Fatty acids and their compounds, the lipids. Reinhold Publishing Corp., New York.

Christie, W.W. (1973). Lipid Analysis. Pergamon Press, Oxford.

Harwood, J(1996). Recent advances in the biosynthesis of plant fatty acids Biochim. Biophys, Acta. 301 : 7-56.

Johnson, A.R. and Davenport, J.B. (1971). Biochemistry and Methodology of Lipids. John Wiley and Sons, Inc., New York.

Kates, M. (1986). Techniques of Lipidology: Isolation, Analysis and Identification of Lipids. 2nd ed., Laboratory Techniques in Biochemistry and Molecular Biology. Vol. 3, Part 2 (Burdon RH, van Knippenberg PH, editors), Elsevier Science Publishing Co., Inc., New York.

Mooke, T. Jr ed (1993). Lipid metabolism in Plants. CRC Press, Boca Raton. FL

Murphy, D.J. (1996). Engineering oil production in rapeseed and other oil crops. Trends Biotechnol 14 : 206-213.

Ohlrogge, J.B. (1994). Design of new plant product engineering of fatty acid metabolism. Plant Physiol. 104 : 821-826.

Varely. H Gowenlock AH and Bell, M (1980). Practical Clinical Biochemisty 5th ed. William Heineman Medical Books Ltd. London.

❑❑❑

Unit -5

Nucleic Acids

Nucleic acids like proteins occur in all living cells. They device their name because of their primary occurrence in the nucleus and acidic nature. Deoxyribose nucleic acid (DNA), one of the two nucleic acids is a major component of chromosomes. Small amounts of DNA are also found associated with chloroplasts and mitochondria.

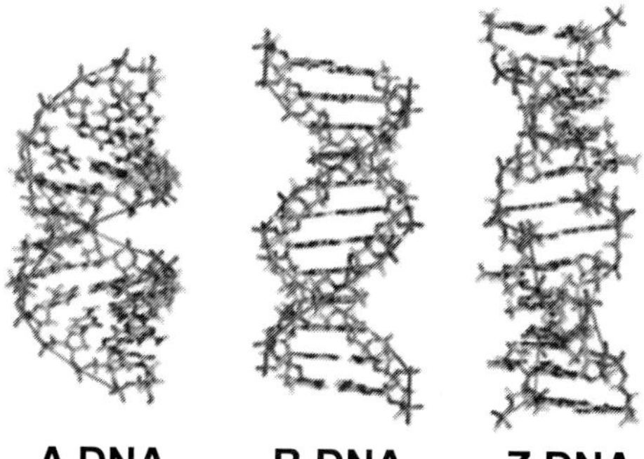

Fig 5.1 : Structure of DNA

Certain animal viruses and many bacteriophages are DNA-protein particles. Ribose nucleic acid (RNA) the other nucleic acid is found mostly in the cytoplasm and in the plant viruses very little amount of DNA and RNA occur in free forms. Nuclear DNA is associated with arginine and lysine rich proteins.

Nucleic acids are defined as "the nucleic acid (DNA and RNA) are the molecular repositories for genetic information and are jointly referred to as the molecules of heredity. The structure of every protein and ultimately of every cell constituent is a product of information programmed into the nucleotide sequence of a cell's nucleic acids".

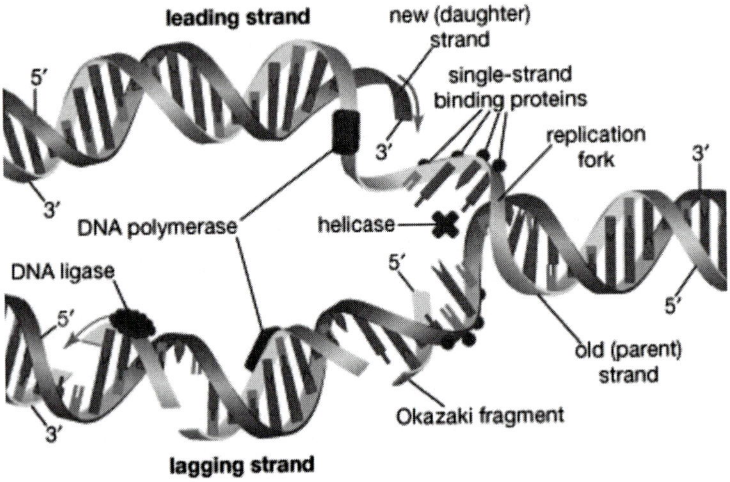

Fig 5.2 : Semiconservative DNA replication

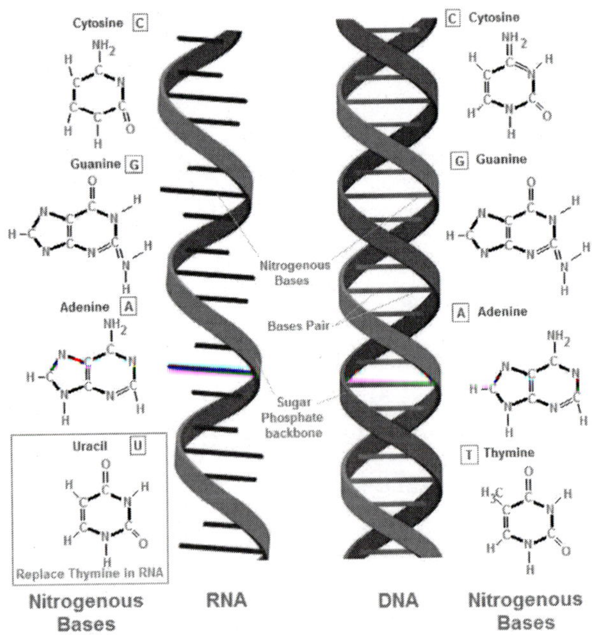

Fig 5.3 : Cell DNA

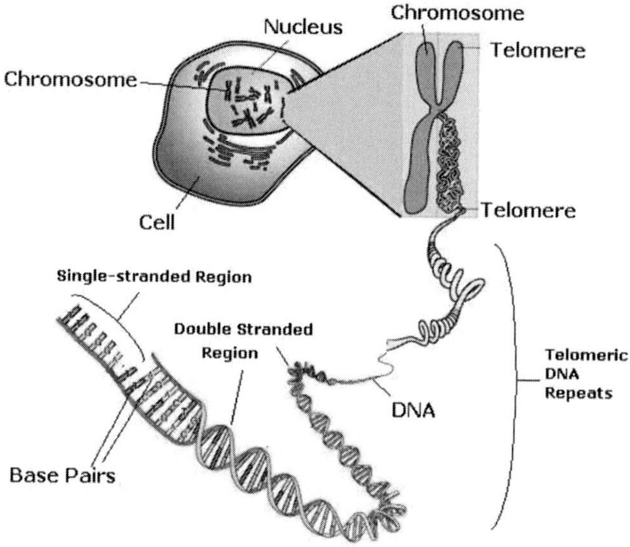

Fig 5.4 : DNA Present in the Nucleus

- *"No amount of experimentation can ever prove me right; a single experiment can prove me wrong."*

  Albert Einstein

- *"Be courteous to all but intimate with few and let those few be well tried before you give them your confidence."*

  George Wahington
  First President of USA

Experiment - 1 : The Determination of the Phosphorus content of a Nucleic Acid

Principle

The nucleic acid is oxidized with perchloric acid to give inorganic phosphate which is then estimated by the usual methods.

Requirements

Reagents

Nucleic acid solution, Regents for the estimation of inorganic phosphate, Perchloric acid (60% v/v).

Glass/Plastic wares: Beaker, measuring cylinder, microtips, Digestion flasks and racks.

Equipments : micropipette, weighing balance.

Procedure

1. Pipette 2 ml of a solution of nucleic acid (0.1 mg/ml) into a digestion flask.
2. Add 0.5 ml of perchloric acid and digest over a low flame for 1 h until the all the inorganic matter has disappeared (*Care*: explosion risk).
3. After cooling, add 1 ml of distilled water to each flask and measure the phosphate.

Observations

....

....

....

....

Experiment – 2 : Estimation of DNA by Diphenylamine (DPA) Method

Practical

This is a general reaction given by deoxypentoses. The 2-deoxyribose of DNA, in the presence of acid, is converted to ω-hydroxyevullin aldehyde which reacts with diphenylamine to form blue coloured complex with absorbance maximum at 600nm. Compounds such as furfural alcohol and arabinol, which can be converted into ω-hydroxyevulinic aldehyde, will also give this reaction. In DNA, since only deoxyribose of purine nucleotides is released, the value obtained represents half of the total deoxyribose in the sample.

Requirements

Reagents

1. Standard DNA solution: Dissolve calf thymus DNA (100 μg/ml) in 1 N $HClO_4$ by heating at 70^0C for 15 min. Make different dilutions of this stock solution ranging from 20-100 μg DNA/ml using 0.5 N $HClO_4$.
2. 1.6% (w/v) acetaldehyde: Prepare by dissolving 1 ml of ice cold acetaldehyde in 50 ml of D/W.
3. Diphenylamine solution: Dissolve 1.5 g of diphenylamine in 100 ml of glacial acetic acid and 1.5 ml of conc. H_2SO_4.
4. Diphenylamine reagent: Prepare by mixing 0.5 ml of 1.6% of acetaldehyde and 100 ml of diphenylamine solutions. This solution must be prepared fresh.

Glassware: Test tubes, conical flasks, Beakers, etc.

Equipments: Spectrophotometer or colorimeter, boiling water bath.

Procedure

1. Take 2.0 ml aliquot of the sample in which DNA has to be estimated in a test tube. In another set of test tubes, pipette 2.0 ml of standard DNA solution of different dilutions. In one of the test tubes take 2.0 ml of 0.5 N $HClO_4$ as a reagent blank.
2. Add 4.0 ml of diphenylamine reagent (reagent no. 5) to all the tubes, mix contents properly and keep at room temperature in dark for 16-18 h or overnight. Alternatively, keep the tubes in

boiling water for 10 min and cool them under running tap water.

3. Record the absorbance at 600 nm.
4. Draw a standard curve of A_{600} vs. DNA concentration. From absorbance of the sample, determine the amount of DNA in it. Express the results as ......mg of DNA/g fresh weight of tissue.

Observations

....

....

....

....

Experiment – 3 : Determination of RNA by Orcinol Method

Principle

This is a general method for estimation of pentoses. Acid hydrolysis of RNA releases ribose which in the presence of strong acid undergoes dehydration to yield furfural. Orcinol, in the presence of ferric chloride as a catalyst, reacts with furfural producing a green coloured compound with absorbance maxima at 665 nm. DNA gives a limited positive reaction with orcinol test.

Requirements

Reagents

1. N-butanol
2. 5% $HClO_4$
3. Standard RNA solution: Dissolve yeast RNA (500 µg/ml) in 5%$HClO_4$. Make different dilutions to obtain solutions containing 100-500 µg RNA/ml with 5% $HClO_4$.
4. Orcinol reagent: Dissolve 100 mg of ferric chloride ($FeCl_3.6H_2O$) in 100 ml of conc. HCl and then add 3.5 ml of 6% solution of orcinol prepared in alcohol.

Glassware: Test tubes, conical flasks, Beakers, etc.

Equipments: Spectrophotometer or colorimeter, boiling water bath.

Procedure

1. Take 2.0 ml solution of each of the dilutions of RNA standard solutions, test sample and 2.0 ml of 5% $HClO_4$, as a blank, in different test tubes.
2. Add 3.0 ml orcinol reagent to all the tubes and mix properly.
3. Keep the test tubes in a boiling water bath for 20 min.
4. After cooling them, add 7.0 ml of n-butanol to each tube and measure the A_{665} against blank.
5. Plot a graph between absorbance at 665 vs amount of RNA and from this standard curve, determine the amount of RNA in the provided sample.

Observations

....

....

....

....

Experiment – 4 : To Perform Restriction Digestion of λ DNA

Principle

rDNA technology is a modern technology in the field of molecular biology. The restriction enzymes cut the DNA at specific site and produce various types of ends. Type II restriction enzymes cleave the DNA within the recognized sequences.

Requirements

Reagents : l DNA, EcoRI, sterile water, agarose, tank, 50 X TAE.

Glassware : Test tubes, conical flasks, Beakers, etc.

Equipments : Spectrophotometer or colorimeter, boiling water bath.

Procedure

1. Take 1 μg of λ DNA in a tube.
2. Add 1 U EcoRI enzyme with 2 μl of buffer.
3. Keep the tube at 37^0C water bath for 2 h.
4. Heat the tube at 65^0C terminate the reaction.
5. Run the digested product on 1% agarose.

Precautions

1. Do not leave enzyme at room temperature.
2. Mix the contents thoroughly.

Observations

....

....

....

....

Experiment – 5 : To Perform Ligation of λ EcoRI Digest using T_4 DNA Ligase

Principle

Construction of recombinant DNA molecule is dependent on the ability to covalently seal single stranded nicks in DNA. This process is accomplished both *in vivo* and *in vitro* by the enzyme DNA ligase. It catalyzes the formation of phosphodiester bonds between justaposed 5′ phosphate and a 3′ hydroxyl terminus of double stranded DNA. It can repair single stranded DNA restriction fragments having either blunt ends or homologous cohesive ends.

T_4 DNA ligase has the unique ability to join sticky and blunt end fragments. Cohesive end ligation is carried out at 12^{0}C to 16^{0}C to maintain a good balance between annealing and activity of the enzyme. If the reaction is set a high temperature, annealing of the ends become difficult, while lower temperature diminishes the ligase activity.

Lack of cohesive termini makes blunt end ligation more complex and significantly slower. So, 10-100 times more enzymes is required to achieve the similar ligation efficiency as that of cohesive and ligation.

Requirements

Reagents: l EcoRI digest, ligase and its buffer, 16^{0}C water bath, pipettes.

Glassware: Test tubes, conical flasks, Beakers, etc.

Equipments: Spectrophotometer or colorimeter, boiling water bath.

Procedure

1. The ligase assay buffer and λ EcoRI digest.
2. Take 1 μl of ligase, 2 μl of buffer, 7 μl of water and 10 μl of λ EcoRI digest.
3. Incubate at 16^{0}C for hr in a pre-set water bath for ligation.
4. Add 2 μl of gel loading buffer to the ligated sample.
5. Pipette 10 μl of λ EcoRI digest (control) and 2 μl of gel loading buffer into a vial, label it as control.
6. Load the ligated and the control sample.
7. Electrophorese the sample at 100V for 1-2 hr.
8. Visualize the DNA bands in UV-transilluminator.

Observations/Interpretation

EcoRI has 5 sites on λ DNA, produces 6 fragments of varied sizes with cohesive ends. By the action of ligase enzyme, the fragments will be joined end to end. Following ligation, the samples will be analyzed wherein the ligated sample will appear as a single band as against 6 bands of λ EcoRI digest.

Six double stranded fragments formed by the digestion of DNA with EcoRI are ligated by T_4 DNA ligase to give a single band.

Experiment – 6 : Elution of Plant DNA Fragment from Agarose Gel

Principle

Freezing and thawing break agarose clumps releasing DNA which upon centrifugation appears in the aqueous phase and gets precipitated with two volumes of cold ethanol.

Requirements

Regents

1. Agarose slab gel containing DNA fragments: Digest the plant DNA with EcoRI and run the digested sample on agarose gel electrophoresis.
2. Tris saturated phenol: Take redistilled phenol and saturate it with 0.1 M Tris. HCl (pH 8.0).
3. Chloroform: isoamyl alcohol (24:1).
4. Tris-saturated butanol.
5. 3 M sodium acetate buffer (pH 5.5).
6. Absolute ethanol.
7. 70% ethanol (v/v).
8. TE

Glassware : Test tubes, conical flasks, Beakers, etc.

Equipments : Spectrophotometer or colorimeter, boiling water bath, knife or a razor blade.

Procedure

1. Cut the agarose gel with a razor blade across the region of your choice.
2. Transfer the gel piece to a centrifuge tube and add Tris-saturated phenol.
3. Freeze at – 70% for minimum of 2h.
4. Thaw out the frozen gel pieces gradually to 4-10^{0}C by keeping the tube in refrigerator.
5. Centrifuge at 4^{0}C at 10,000 rpm for 10 min.
6. Remove the aqueous layer and add equal volume of chloroform: isoamyl alcohol to it. Centrifuge as above in Step 5.

7. Collect the upper aqueous phase and add equal volume of Tris-saturated butanol to remove ethidium bromide. Centrifuge as above and collect the lower layer.
8. Add 3 M sodium acetate, pH 5.5 in such a way that its concentration in the solution is 0.25 M.
9. Precipitate DNA by adding two volumes of chilled absolute ethanol. Keep overnight at - 20^0C.
10. Centrifuge at 4^0C at 10,000 rpm for 10 min. Wash the precipitates once with 70% ethanol, air dry and dissolve in suitable volume (10µl) of TE buffer. Perform 1% agarose gel electrophoresis.

Observations

....

....

....

....

Experiment – 7 : Isolation of total RNA

Principle

Total RNA isolation is based on using guanidine thiocyanate phenol:choloroform procedure of Chomoczynski and Sacchi. Samples are lysed in highly denaturing condition inactivating RNases, followed by phenol: chlroform extraction where DNA and proteins are removed and then precipitated with isopropanol. An additional step of suspending and reprecipitating ensures further purity of the RNA.

Requirements

Reagents : Bacterial cell (10^6), Water saturated phenol, chloroform: isoamyl alcohol(49:1),DEPC water, 1X TE buffer, denaturing solution, isopropanol, 75% ethanol (prepared with DEPC treated water, 0.1% v/v)

Glassware : Test tubes, conical flasks, Beakers, etc.

Equipments : Spectrophotometer or colorimeter, boiling water bath.

Procedure

1. Homogenize the sample in 1 ml of denaturing solution.
2. Add 1 ml of water saturated phenol followed by 200 μl of chloroform: isoamyl alcohol mix (freshly prepared 49:1).
3. Mix thoroughly and incubate on ice for 15 min.
4. Centrifuge at 10,000 rpm for 20 min at 4^0C.
5. Transfer upper aqueous phase carefully to another tube.
6. Add 1 ml of isopropanol to precipitate RNA. Incubate at -20^0C for 30 min.
7. Centrifuge at 10,000 rpm for 20 in at 4^0C, discard the supernatant.
8. Resuspend the pellet in 0.3 ml of denaturing solution, precipitate RNA by adding 0.3 ml of 100% isopropanol, incubate at -20^0C for 30 min, centrifuge at 10,000 rpm for 20 min at 4^0C and discard the supernatant.
9. Resuspend the RNA pellet in 75% ethanol. Incubate at RT for 10-15 min to dissolve residual amounts of guanidine.

10. Centrifuge at 10,000 rpm for 20 min at 4^0C, discard the supernatant.
11. Do not over dry the pellet. Dissolve in 100-200 μl of DEPC water. Incubate for 10-15 min at 55^0C to ensure complete solubilization of RNA.
12. Store the total RNA at -70^0C.

Precautions

1. The RNase should be removed completely. It is very important to inactivate RNase as quickly as possible and minimize contamination. They are stable and do not require any cofactors to function. Therefore small amount of RNase in an RNA preparation will create a real problem.
2. Water or any salt solution used should be treated with Diethyl pyrocarbonate (DEPC) and then autoclaved. This chemical inactivates ribonuclease by covalent modification. Solutions containing Tris cannot be effectively treated with DEPC because Tris reacts with DEPC to inactivate it.
3. Glassware and plasticware should be treated to get rid of any RNase but autoclaving will not fully inactivate RNase. Glassware can be baked at 300^0C for 4h. Plastic ware can be rinsed with chloroform to inactivate RNase.
4. Wear gloves while handling as hands are major sources of contamination.
5. Clean the working are with 70% ethanol before use.
6. Water saturated phenol is supplied with aqueous layer. When pipetting, be sure not to take upper aqueous layer.

Observations

....

....

....

....

Experiment – 8 : To Perform Southern Hybridization

Principle

Southern Hybridization technique involves transfer of DNA fragments separated in electrophoretic gels to membrane filters for detection of specific base sequences by complementary probes. Prof E.M. Southern developed this technique in 1975, hence referred as Southern transfer. Southern blots are used to identify and quantitate specific DNA sequences, in analysis of genome organization and expression, in the study of genetic diseases, in DNA fingerprinting and analysis of PCR products.

DNA molecules are resolved on the gel on the basis of their size and transferred to positive charged nylon membrane by capillary or electrophoretic transfer. The DNA is immobilized onto the membrane by UV crosslinking. The membrane is then incubated with the protein block to reduce non-specific interaction. Hybridized probe is detected by incubating the membrane with streptavidin-HRP (Horse radish peroxidase) conjugate and finally incubating with the substrate solution until sufficient colour develops. The final detection is done following addition of substrate, TMB/H_2O_2 (Tetramethyl benzidine H_2O_2-substrate) that reacts with HRP to give a blue coloured DNA band on the nylon membrane.

In brief, to learn the technique of Southern Hybridization involves the following Practicals:

1. Electrophoretic separation of DNA molecules by agarose gel electrophoresis
2. Electrophoretic transfer of DNA from agarose gel to nylon membrane
3. Immobilization of DNA on to nylon membrane
4. Hybridization and non isotope detection of DNA of interest

Requirements

Reagents

Ice, DNA marker, Biotinylated probe, Prehybridization buffer, Hybridization buffer, Blocking Buffer, 2X wash buffer (A, B, C & D), Streptavidin HRP conjugate, Conjugate dilution buffer, 10X substrate, 50 X TAE, Blocking Powder, Tween-20, agarose, Filter paper, Nylon Membrane, petridish, hot air oven, Incubator shaker (45^0C), UV-transilluminator, distilled water, ethidium bromide.

Glassware: Test tubes, conical flasks, Beakers, gloves, measuring cylinders, tips, transparent polythene sheet.

Equipments: Spectrophotometer or colorimeter, boiling water bath, thermometer, micropipette, blade, scissors, forceps.

Solutions

1. Preparation of Prehybridization buffer: Add 1 g of blocking powder to 10 ml of prehybridization buffer.
2. Preparation of Hybridization buffer: Add 1 g of blocking powder to 10 ml of hybridization buffer.
3. Preparation of blocking buffer: Add 1 g of blocking powder to 10 ml of blocking buffer.
4. Preparation of wash buffers A, B, C & D buffer: Thaw and mix the buffer provided thoroughly. Dilute 15 ml each of 2 X buffers supplied with 15 ml of distilled water to give buffers at 1 conc. 30 ml each.
5. Preparation of Streptavidin HRP-conjugate buffer: Prepare 9 ml of conjugate dilution buffer by adding 9µl of Tween-20 to the conjugate dilution buffer. Dilute 3 µl of Streptavidin-HRP conjugate with 9 ml of conjugate dilution buffer for each practical just before use.
6. Preparation of substrate buffer: Dilute 0.5 ml of substrate solution with 4.5 ml of distilled water to give a final volume of 5 ml substrate solution of 1 X concentration.
7. Preparation of electrotransfer buffer: Dilute 25 ml of 10 x electrotransfer buffer 225ml distilled water to give a final vol of 250 ml electrotransfer buffer at 1X conc.

Procedure

Day 1: Agarose Gel Electrophoresis

1. Prepare 1.0% agarose gel containing ethidium bromide.
2. Load 20 µl of ready to use DNA marker supplied onto the gel. Run at 50-100 V until the dye reaches 4.5 cm from the well.
3. Cut the DNA marker lane from the agarose gel as follows: Cut the gel ~ 3 mm above the first band and ~ 2 mm below the last

band, ensuring the gel measures about 4 to 4.5 cm. Mark the appropriate position of the gel to be cut on the gel tank.

4. Cut the filter paper (2 nos.) and nylon membrane exactly to the size of the cut gel. Ensure there is no protrusion of the filter paper and the membrane from the gel.
5. Wet the cut gel, nylon membrane, filter papers and the electrotransfer cassette in 1x electrotransfer buffer for electroblotting.

Electroblotting

Assemble the electrotransfer apparatus. Start the arrangement by placing filter paper on the cathode cassette cover followed by the cut gel and nylon membrane. Mark and place the soft of the nylon membrane to the cut gel. Place the wet filter paper on the nylon membrane followed by anode cassette cover. Tighten the electrotransfer cassette tightly with the screws.

6. Insert the cassette into the apparatus filled with 250 ml of 1X electrotransfer buffer.
7. Connect the cords to power supply and set the voltage to 50 V for 3.5 hr.
8. Turn off the power supply and remove the cassette from the apparatus. Drain the buffer.

Immobilization of DNA on Membrane

9. Remove the nylon membrane gently from the cassette and place it on a thin transparent polythene sheet and place this on a UV transilluminator (expose the soft side of membrane containing transferred DNA to UV light) with UV lamps switched on for 20 min. This helps in fixing the DNA on the membrane.
10. Turn of the UV transilluminator. Place the membrane in plastic petridish provided and incubate in a hot air oven at 70^0C for 30 min. This ensures complete immobilization of DNA on to the membrane.

Hybridization

11. Bring the petridish containing the membrane to room temperature after incubation. Add 10 ml prehybridization

buffer to the petridish and incubate at 45^{0}C incubator shaker with mild shaking (70-90rpm) for 45 min.

12. After incubation, discard prehybridization buffer taking care not to discard the membrane.
13. Add 10 ml hybridization buffer to the petridish containing the membrane.
14. Keep 1 vial of biotinylated probe for 10 min in a boiling water bath and immediately chill by placing it on ice for 5-10 min. Add this probe to the hybridization buffer in the petridish. (Rinse the probe vial with 300μl of hybridization buffer and add it to the petridish). Incubate the petridish at 45^{0}C incubator shaker with mild shaking 70rpm for 16 hr.

Day 2

15. Decant the hybridization buffer, add 10ml of 1X wash buffer A and gently swirl the petridish for 5 min at RT. Repeat the washes twice, each wash of 5 min. Discard the buffer after each wash.
16. Add 10 ml of 1 X prewarmed (70^{0}C) wash buffer B and gently swirl the petridish. Incubate at 70^{0}C for 5 min in a hot air oven and gently swirl. Repeat the washes or another two times. Discard the buffer after each wash.
17. Add 10 ml of 1 X blocking buffer to the petridish and incubate at room temperature for 1h with gentle rocking.
18. Discard the blocking buffer.
19. Add 9 ml of diluted HRP-streptavidin conjugate to the petridish and incubate at RT for 20 min with gentle rocking. Discard the conjugate buffer.
20. Add 10 ml of 1 X wash buffer C to the petridish and incubate at RT for 5 min with gentle rocking. Repeat the washes two more times. Discard the buffer after each wash.
21. Add 10 ml of 1 X wash buffer D to the petridish and incubate at RT for 5 min with gentle rocking. Repeat the washes two more times. Discard the buffer after each wash.
22. Add 5 ml of 1 X substrate solution and gently swirl at RT for 15-20 min until a blue colour band develops.

23. After blue colour band is seen, stop the reaction by placing the membrane in distilled water.

Precautions

1. Carry out the Practical by wearing gloves. Specially while handling ethidium bromide agarose gel.
2. Agarose gel to be prepared by addition of ethidium bromide at a conc. of 0.5μg/ml from a stock of 10mg/ml.
3. Dilute the appropriate quantities of buffers (Wash buffers A, B, C & D) to 1X conc. with distilled water just before use.
4. Dilute 10 X electrotransfer buffer to 1X with distilled water.
5. Use autoclave tips and measuring cylinders to avoid contamination.
6. All the buffers provided should be brought to room temperature one day prior to use except streptavidin HRP-conjugate buffer and substrate solution.
7. Buffers can be prepared according to the required conc. on the previous day of use except for streptavidin HRP-conjugate buffer and substrate solution (Which have to be prepared just before use).
8. Ensure no air bubbles are present between any of the layer of filter paper, cut gel and the nylon membrane.

Observation

Hybridization is a technique to identify the DNA of interest from the pool of DNA fragments. In this nonisotopic detection technique, a biotinylated probe (sequence complementary to the target DNA) is used. The probe binds to the complementary sequences of the DNA marker. This hybridized complex (probe-target complex) is detected by streptavidin-HRP conjugate, an enzyme that binds to biotin molecule of the probe. On addition of the substrate solution, the enzyme reacts with the substrate to form the blue coloured band. The probe-target complex is seen as the major blue band however 1 or 2 very less intense blue band may be seen due to random non specific annealing of the probe to the DNA bound on the membrane.

Experiment – 9 : To Perform PCR Amplification of Specific Target Sequence from Genomic DNA

Principle

PCR, polymerase chain reaction is an *in vitro* method of enzymatic synthesis of specific DNA sequence, developed by Karry Mulis in 1983. It is a very simple and inexpensive technique for characterizing, analyzing and synthesizing any specific piece of DNA/RNA form virtually any living organism plant/ animal/ virus/ bacteria PCR is also known as Molecular Photocopying.

PCR consist of 3 basic steps:-

1. **Denaturation:** During this step, the two strands melt open to form Ss DNA and all enzymatic reactions stop. This is generally carried out at 92^0C-96^0C.
2. **Annealing:** Annealing of primers to each original strand for new synthesis is carried out between 45^0C-55^0C.
3. **Extension at 72^0C:** The polymerase adds NTPs complementary to template at 3′ ends of the primers. Since bout strands are copied during PCR, there is an exponential increase in number of copies of the genes.

These steps are repeated 20-30 times is an automated thermocycler that can heat & cool the reaction mixture in tubes within very short time.

$$\text{Amplification} = 2^n$$

Where, n = number of cycles

Procedure

1. The reagent were added in PCR tubes in following order-

Sterile water	38 μl
10 X assay buffer	5 μl
10 mm dNTP mixture	3 μl
Template DNA (100 ng/μl)	1 μl
Forward primer (10 ng/μl)	1 μl
Reverse primer (100 ng/μl)	1 μl
Taq DNA polymerase (3 u/μl)	1 μl
Total reaction volume	**50 μl**

2. The amplification was carried out in a thermocycler for 30 cycles using the following reaction condition.
   (a) Initial denaturation was carried out for 1 min. at 94^0C.
   (b) Denaturation was carried out for 30 sec. at 94^0C.
   (c) Annealing was carried out for 30 sec. at 48^0C.
   (d) Extension was carried out for 1 min. at 72^0C.
   (e) Final Extension was carried out for 2 min. at 72^0C.
3. The gel was visualized under UV transilluminator.

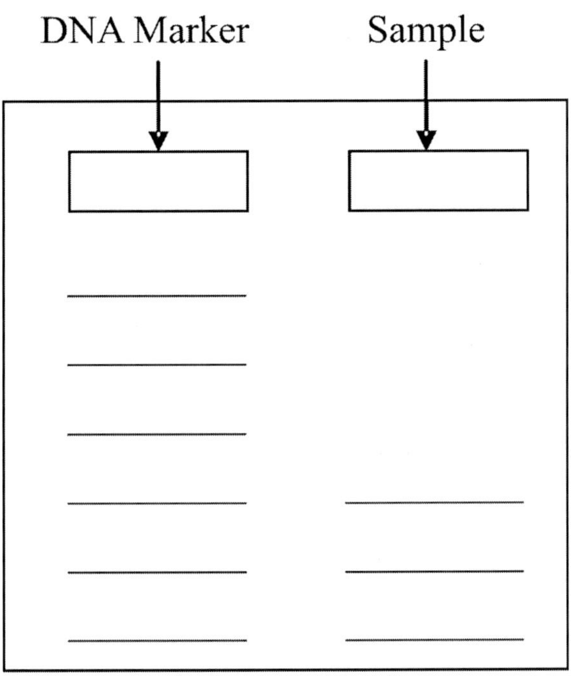

Observations

....

....

....

....

Experiment – 10 : To Isolate the total Genomic DNA from Plant Source

Principle

For the extraction of total genomic DNA from plant sample the plant cell were typed by dehydration of cell wall protein by proteolytic enzyme/agent. The released genomic DNA is protected from DNA are by inhibiting the activity. The most suitable detergent used for lysis of plant cell used i.e. CTAB or SDS. The isolated DNA is then precipitated using alcohol. The most suitable method for precipitating nucleic acid is with the use of Ethanol or isopropanol. Alcohol precipitation is valuable because they allow the nucleic acid to be soluble pellet. The pellet then can be suspended in a buffer of choice. The success of peptization increased by adding optimum concentration of non-evalent cation present to the high negative charge of phosphate present in nucleic acid without this repulsion is to great for the nucleic acid to aggregate and form a precipitate. Extraction of genomic DNA occurs in 3 stages-

(i) Homogenise 1 g of leaves with liquid N_2 and mix with 2 ml of extraction buffer + 6 μl of β-mercaptoethanol.

(ii) Deproteinizing the extract using phase separation.

(iii) Alcohol precipitation of Nucleic Acid.

Requirements

Reagents

1. Liquid N_2 or dry ice
2. Extraction buffer

   2% CTAB, 20 mM EDTA, 2% v/v PVP (Polyvinyl Pyrolidol), 5 mM Ascorbic Acid, 1.4 M NaCl, 100 mM Tristice (pH=8) = 100 mM
3. β-Mercaptoethanol: 6 μl
4. Tris EDTA (HCl): pH = 8-9, bring pH = 7, EDTA = 0.74 gm Dissolve in distilled water (100 ml) + Tris, HCl = 15.7 gm
5. Chloroform: Isoamyl Alcohol (24:1)
6. Chilled Alcohol (90%): Isopropyl Alcohol
7. 70% Ethanol Solution

Glassware : Test tubes, conical flasks, Beakers, gloves, measuring cylinders, tips, transparent polythene sheet.

Equipments: Centrifuge, Incubator, boiling water bath, micropipette, scissors, forceps.

Procedure

1. Homogenise 1 g of leaves with liquid N_2 and mix with 2 ml of extraction buffer + 6 µl of β-mercaptoethanol.
2. Incubate at 60^0 C for 45-60 min. on water bath, give occasional gentle, inversion on tubes.
3. After Incubation, add 100 µl of solution of chloroform + Isoamyl alcohol (24:1).
4. Mix both the phases by gently inversion till the formation of emulsion.
5. Centrifuge at 10000 rpm for 15 min.
6. Pipette out upper aqueous phase in a fresh microfuge tube.
7. Add equal amount of chilled Isoproponal and mix by gentle inversion.
8. Incubate for 2 hrs at – 20^0C.
9. After incubation centrifuge at 13,000 rpm for 15 min at 4^0C.
10. Discard the supernatant and add 200 µl of chilled 70% Ethanol.
11. Incubate for 15 min. at room temp. and centrifuge at 13,000 rpm for 15 min.
12. This step can be repeated 2-3 times if required.
13. Discard the supernatant and air dry the pellet in the thermoset.
14. Add 10 µl of TE buffer.
15. Visualize through agarose gel Electrophoresis.

Observations

....

....

....

....

Additional Readings

Boyer, R.F. (1993). Modern Experimental Biochemistry 2nd ed. The Benjamin/Cummings Publishing Company Inc. California U.S.A.

Brooker, R.J. (2005). Genetics Analysis and Principles, 2nd edn. New York: McGraw-Hill.

Britt, AB(1996). DNA damage and repair in plants Annu. Rev. Plant Physiology. Plant Mol Biol. 47 : 75-100.

Cooper, T.C. (1997). Tools of Biochemistry. John Wiley, NY. USA.

Davidson, J.N. (1976). The biochemistry of the Nucleic Acids. 8th ed., Chapman and Hall, London.

Holme, DJ and Peck, H (1998). Analytical Biochemistry. 3rd ed. Addison Wesely Longman Ltd. Essex. U.K.

Lodish, H. et al. (2008). Molecular Cell Biology, 6th edn. San Francisco, CA: W.H. Freeman.

Sugiura. M (1992). The chloroplast genome. Plant Mol. Biol. 19 : 149-168.

Tjian, R(1995). Molecular machines that control genes, Scientific Am. 272 (2) : 54 -61.

Wang J.C (1996). DNA topoisomerases. Annu. Rev. Biochem. 65 : 635-692.

Walker, J.M. and Raply, R. (2008). Molecular biomethods handbook, 2nd edn. Totowa, NJ: Human Press.

Younghusband, H.B. and Inman, R.B. (1974). The electron microscopy of DNA. Ann. Rev. Biochem. 43: 605.

❑❑❑

Unit -6

Enzymes

The term enzyme was coined in 1878 by Friedrich Wilhelm Kuhne to designate these biological catalysts that had previously been called ferments. As they quicken most of the chemical reactions occurring in the body, the enzymes have been designated as the "manifestations of nature's impatience".

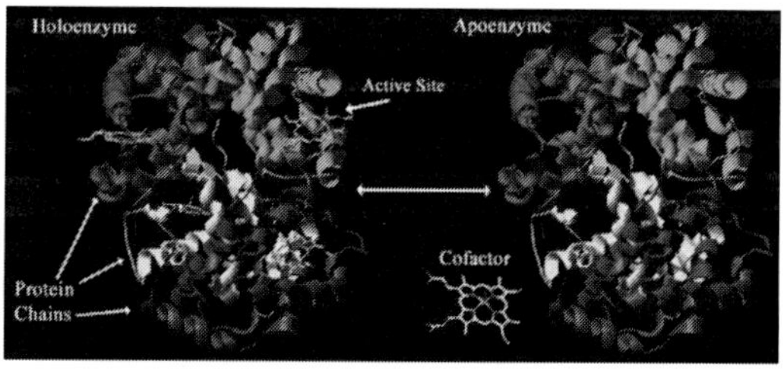

Fig 6.1 : Enzyme Structure

The name enzyme (en[G] = in; zyme[G] = yeast) literally means 'in yeast'. This was referred to denote one of the most noteworthy reactions wherein the production of ethyl alcohol and carbon dioxide through the agency of an enzyme, the zymase, present in yeast takes place. This reaction is most popularly known as alcoholic fermentation. Sumner and Myrback (1950) have beautifully defined the enzymes as

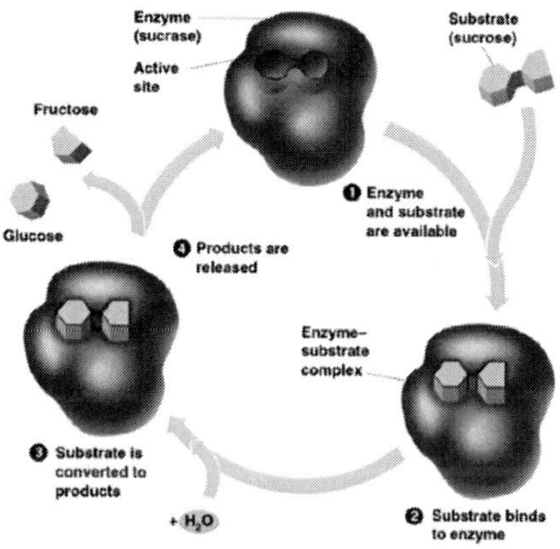

Fig 6.2 : The catalytic cycle of an Enzyme

'simple or combined proteins acting as specific catalysts'. They are soluble, colloidal molecules which are produced by living cells. All enzymes are globular proteins with a complex 3-'D' structure, capable of binding substrate molecules to a part of their surface. They affect the life of an organism to such an extent that life has aptly been called as "an orderly function of enzymes".

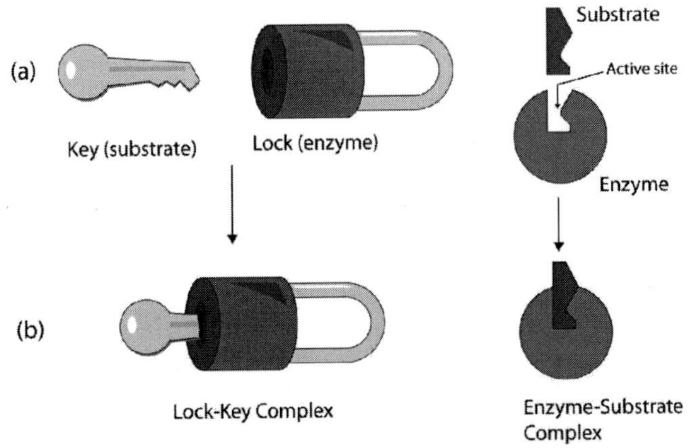

Fig 6.3 : Lock-Key Model

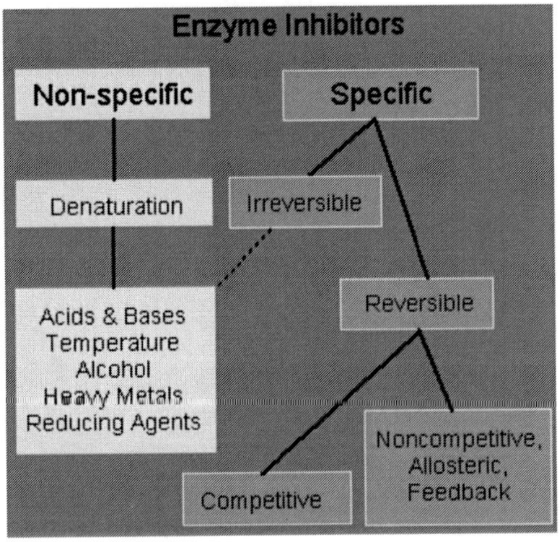

Fig 6.4 : Enzyme Inhibitors

- *"Science is a first-rate piece of furniture for a man's upper chamber, if he has common sense on the ground floor."*

Oliver Wendell Holmes

Experiment – 1 : Determination of Catalytic Power of Acid Phosphatase

Principle

Enzymes are the major class of proteins, called biological catalysis or biocatalysts because the speed up the rate of reaction occurring inside the cells of all organisms. In our body, numerous biological reactions occur which are not possible without enzymes. The term enzymes was coined 1887 by Friedrick W. Kuhne to designate the biological catalysts that had earlier been called ferment acid phosphatase is located in lysosomes having acidic pH falling between 5 and 6. The pH of cytoplasm is nearly neutral. When an enzyme is active only in acidic pH, it is called acid-phosphatase.

The enzyme catalyzes the release of inorganic phosphate from the compound of organophosphates i.e. nucleotides, nucleic acid, phosphorylated proteins, etc. enzyme activity is assayed by measuring the conversion of substrate into products which is formed within a given time. For determining the activity of acid phosphatase an artificial substrate, *p*-nitrophenyl phosphate, is used which is broken by acid phosphatase as below:

$$p\text{-nitrophenyl phosphate} + \text{water} \xleftrightarrow{\text{Enzymes}} p\text{-nitrophenol} + \text{inorganic phosphate (Pi)}$$

(yellow colour)

p-nitrophenyl phosphate (PNPP) is a colorless substrate but after reaction it imparts yellow coloured due to liberation of *p*-nitrophenol which can be easily measured is a colorimeter. The colour intensity can be correlated with enzyme activity by using a standard graph. Enzyme fraction is procured from any source as aqueous solution. It is mixed with PNPP in a suitable buffer at 5.7 pH and allow to react. At the end of the reaction an aliquot of potassium hydroxide is added which stops the reaction and inactivates the enzyme. Besides, this alkali intensifies the yellow colour through favouring the resonance of PNP. This facilitates to detect the colour change in spite of the presence of small amount of PNP.

Requirements

Sprouted seed of mung dal (legume),

Reagents: Acetate buffer (1M, pH 5.7), *p*-nitrophenyl phosphate (0.05 M) solution, *p*-nitrophenol (200 mM solution) solution, Potassium hydroxide (0.5M solution, 500 ml), Sodium acetate, acetic acid, sodium chloride (0.9%, saline solution).

Glass/Plastic wares: beaker, Test tubes, mortar and pestle, glass pipettes (1 ml, 5 ml), cheese clothes.

Equipments: pestle and mortar, cheese cloth, refrigerator, colorimeter/ spectrophotometer

Procedure

1. Take mung seeds and soak overnight to allow germination under dark condition.
2. Weigh 10 g of sprouted seeds, wash with distilled water (5 times) in a beaker.
3. Place pestle and mortar, transfer small amount of sprouted seeds and pour a few volume of saline solution. Care should be taken that the total amount of ground material should be about 50 ml.
4. Take cheese cloth, make 5 layers of it and squeeze the ground material.
5. Collect the filtrate in a test tube. This filtrate consists of the required enzyme. Since room temperature fluctuates during different seasons, the enzyme fraction should be place in a freeze till further use.

Acetate buffer	=	0.4 ml
PNPP solution	=	0.2 ml
Distilled water	=	0.2 ml
Enzyme fraction	=	0.2 ml

(Immediately after adding enzyme fraction (0.2 ml), mix the contents gently and note the time. Cover the test tubes and place in the dark for 20 – 25 minute incubation at 25 ± 1^0C)

6. After 20 minutes of incubation, add 3 ml of KOH solution with the help of sterile autopipette or using a pipette sucker on the mouth of a pipette (**mouth pipetting of KOH is very dangerous**).
7. Mix the contents properly by gentle shaking of tube.

8. Take 1 to 6 tubes and mark it properly. Pour the reagents in the order as given below.

Tube No.	Distilled Water (ml)	PNP Solution (ml)	KOH (ml)
1. (control)	0.1	0	3
2.	0.8	0.2	3
3.	0.6	0.4	3
4.	0.4	0.6	3
5.	0.2	0.8	3
6.	1.0	1.0	3

9. Mix the contents properly and then measure absorbance of each solution at 405 nm of wavelength by using a colorimeter/ spectrophotometer. Tube No. 1 is used as blank for setting the absorbance of spectrophotometer initially at zero.

Results

Record the absorbance serially from 1 to 6. The solution of PNP is 200 mM it means 1 ml of solution contains 0.2 m mol of POP. Record absorbance as below:

Tube no.	PNP amount (m mol)	Absorbance
1.	0	0
2.	0.04	
3.	0.08	
4.	0.12	
5.	0.16	
6.	0.2	
Assay Tube 1	?	
Assay Tube 2	?	
Assay Tube 3	?	

Plot a standard graph between the amount of PNP versus absorbance. Repeat the activity as described earlier for preparation of standard graph for protein estimation. Using this graph determine the amount of PNP comparing with absorbance of known substrate. Report the enzyme activity as micromolecules PNP formed per minute. The enzyme assay was done for 20 minute, so for estimation the amount of PNP formed, divide the amount by 20.

Experiment – 2 : To Obtain the Progress Curve During the Hydrolysis of p-Nitrophenyl Phosphate by Serum Alkaline Phosphatase *(Orthophosphoric Monoester Phosphohydrolase,* 3.1.3.1)

Principle

Alkaline phophatase (pH 9-10 optimum) is found in bone, kidney, liver and intestine and acts on phosphoric esters with the liberation of inorganic phosphate.

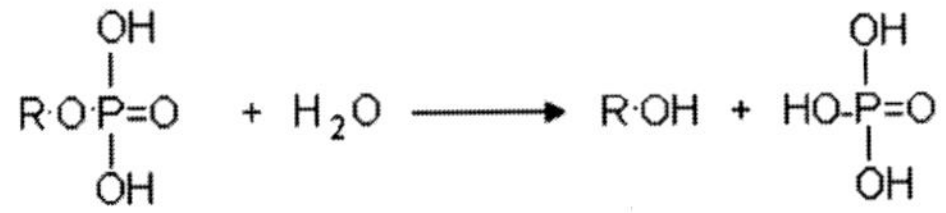

p-Nitrophenyl phosphate is used as substrate and the *p*-nitrophenol released by enzymic hydrolysis is measured colorimetrically. In alkaline solution, *p*-nitrophenol absorbs at 405 nm. The substrate *p*-nitrophenyl phosphate does not absorb at this wavelength so the progress of the enzyme-catalysed reaction can be readily followed by measuring the change in extinction at 405 nm.

Requirements

Serum (50 ml)

Reagents

1. 500 ml sodium carbonate-bicarbonate buffer (0.1 mol/l, pH 10.0).
2. 500 ml *p*-Nitrophenyl phosphate substrate solution. (5 mmol/l in the alkaline buffer).
3. 1l *p*-Nitrophenol standard (50 μmol/l). This is prepared by dissolved 69.6 mg *p*-nitrophenol in 100 ml of alkaline buffer, then diluting this solution one in a hundred with the buffer. Solution (2) and (3) should be freshly prepared.

Glass/Plastic wares: Beakers, Test tubes, micro tips

Equipments: Colorimeter, micropipetes

Procedure

1. ***Test***: Add 0.8 ml of the carbonate-bicarbonate buffer to 2 ml of the substrate solution and mix thoroughly. At zero time, add 0.2 ml of serum and follow the change in extinction at 405 nm.

2. ***Blank*:** This is the same as the test except that 0.2 ml of buffer replaces the serum.

3. ***Standard*:** Prepare a range of *p*-nitrophenol solutions and plot a graph of extinction at 405 nm against concentration.

Calculation

Subtract the extinction of the blank form that of the test and calculate the amount of *p*-nitrophenol released from the standard curve. Express the activity of the serum alkaline phosphate in terms of units per milliliter of serum remembering that 0.2 ml is used in the assay.

Enzyme Concentration

Enzyme assays are carried out on an aliquot of the material available and the activity referred back to the total volume of solution or weight of the enzyme. For example, in clinical biochemistry, only 0.1 ml or 0.2 ml of serum is used and the biochemistry, only 0.1 ml or 0.2 ml of serum is used and the activity expressed in units per litre. For this reason it is important to establish that the activity varies linearly with enzyme concentration. Any deviation from linearity is the result of some limiting factor.

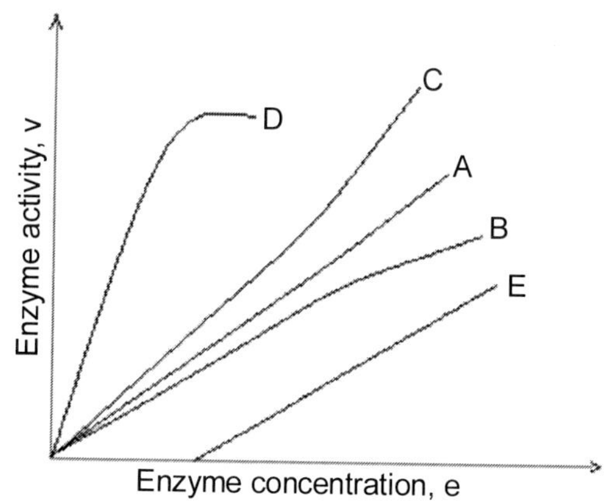

A. Normal response

B. (i) Non-linear part of progress curve used to calculate activity.
(ii) Inhibitor in enzyme preparation.

C. Activator in enzyme preparation

D. Substrate exhausion.

E. Inactivator such as a heavy metal in the reaction mixture.

The effect of enzyme concentration on activity

Observations

....

....

....

....

Experiment – 3 : Variation in Serum Alkaline Phosphatase Activity with Enzyme Concentration

Principle

Phosphatases liberate inorganic phosphate from organic phosphate esters. Acid phosphatase (3.1.3.2) hydrolyzes a number of acid phosphomonoesters and phosphoproteins. Alkaline phosphatase catalyses the hydrolysis of numerous phosphate esters, such as esters of primary and secondary alcohols sugar alcohols, phenols and amines. Phosphodiesters are not hydrolyzed by either of them. The enzyme phosphatase hydrolyzes p-nitrophenol phosphate. The released p-nitrophenol is yellow in colour in alkaline medium and is measured at 405 nm. The optimum pH for acid and alkaline phosphatases are 5.3 and 10.5, respectively.

Requirements

Prepare 21 of buffer and substrate and 200 ml of serum as in the previous experiment. This is sufficient material for 50 pairs of students.

Procedure

Repeat the previous experiment with volumes of serum form 0.1 to 0.6 ml. the volume of substrate is kept constant at 2 ml and the volume of buffer adjusted so that the final mixture is 3 ml as previously. A progress curve for each volume of serum is prepared as previously described and the initial reaction velocity is determined. A graph is then plotted of enzyme activity and enzyme concentration expressed as milliliters of serum.

Choose a suitable time which is beyond the linear portion of the progress curve and express the activity as the average amount of *p*-nitrophenol liberated per minute. Plot the activity against enzyme concentration and compare the result with the previous curve, when initial rates were used to express the activity.

Enzyme Activity and Substrate Concentration

Michaelis-Menten enzymes

Kinetics: If the activity of an enzyme is determined over a range of substrate concentrations a curve similar to the rectangular hyperbola.

At low substrate concentration v varies linearly with s, giving *first-order kinetics*: $v = -ds/dt = ks$, where k is the rate constant.

At high substrate concentrations v is independent of s, giving *zero-order kinetics*: $v = -ds/dt = \text{constant} = V$.

At intermediate substrate concentrations there is a mixture of first- and zero-order kinetics.

An equation relating v and s can be obtained for the whole curve and was first derived by Michaelis and Menten in 1913:

$$v = \frac{Vs}{s + K_m}$$

Where V and K_m are constants.

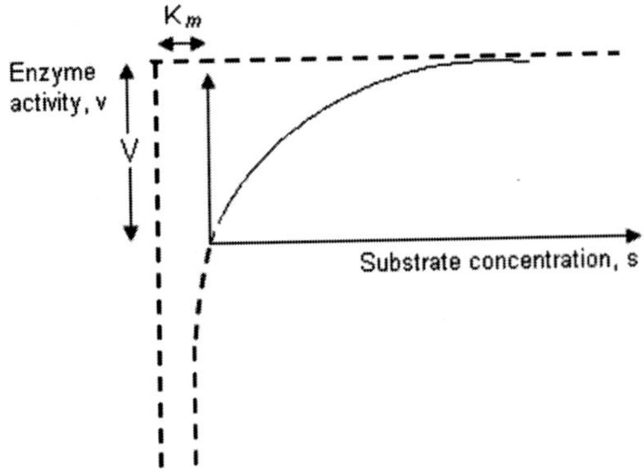

The effect of substrate concentration on enzyme activity

The basic assumption was that the enzyme and substrate form a complex which then breaks down to give the enzyme and products.

Derivation of the michaelis equation: In deriving this equation, Michaelis and Menten assumed that the rate of breakdown of the complex was small enough not to disturb significantly the equilibrium between the enzyme and the substrate. Briggs and Haldane later extended this idea and derived an equation assuming steady-state conditions, i.e. the rate of breakdown of the complex is the same as the rate of formation during the period of the measurement. This more general assumption is made in the following derivation.

The whole process can be represented by the general equation:

$$\underset{(e-p)}{E} + \underset{(s)}{S} \underset{k_{-1}}{\overset{k_{+1}}{\rightleftharpoons}} \underset{(p)}{ES} \xrightarrow{k_{+2}} E + \text{Products},$$

Where e = enzyme (E) concentration, s = substrate (S) concentration, p = enzyme-substrate complex (ES) concentration, k_{+1}, k_{+2}, k_{-1} are velocity constants. The concentration of the free substrate is taken to be the same as the total, since the amount actually bound in the complex will usually be small.

The rate of formation of ES = k_{+1} (e – p)s.

The rate of breakdown of ES = k_{-1} p + k_{+2} p.

Under steady-state conditions, the rate of formation = the rate of breakdown so:

$$K_{+1}(e - p)s = (k_{-1} + k_{+1})p.$$

Rearrangement of this equation gives:

$$p = \frac{es}{s + (k_{-1} + k_{+2})/(k_{+1})}$$

The velocity of the enzyme reaction is given by $v = k_{+2}$ p, so:

$$v = \frac{k_{+2}es}{s + (k_{-1} + k_{+2})/(k_{+1})}$$

When the entire enzyme is present as the complex, then this is the maximum velocity possible and is given by V = k_{+2} e. If the expression $(k_{-1} + k_{+2}) / (k_{+1})$ is replaced by K_m, then the above equation becomes:

$$v = \frac{Vs}{s + K_m} \text{ or } v = \frac{V}{1 + K_{m/s}}$$

This can be rearranged to give the form usually shown for the equation of a rectangular hyperbola:

$$(V\text{-}v)(K_m + s) = VK_m$$

The Kinetic constants K_m and V. If $s = K_m$, then $v = V/2$ so that the Michaelis constant is the substrate concentration which gives half the maximum velocity.

If the equilibrium conditions suggested by Michaelis and Menten hold, then k_{+2} is small compared with k_{+2} and can be ignored, so that $K_m = k / k_{+1}$, which is the dissociation constant for the enzyme-substrate complex. A large K_m therefore means a large dissociation constant or a small association constant ($1/K_m$). Conversely a small K_m means a small dissociation constant or a large association constant ($1/K_m$). The Michaelis constant therefore gives a measure of the enzyme-substrate affinity.

Large K_m = low enzyme – substrate affinity.

Small K_m = high enzyme – substrate affinity.

The kinetic constants K_m and V are most conveniently determined from a linear transformation of the Michaelis equation, obtained by taking reciprocals:

$$\frac{1}{v} = \frac{1}{V} + \frac{K_m}{V} \times \frac{1}{s}$$

A plot of $1/v$ against $1/s$ therefore gives a straight line of slope K_m / V. The reciprocals of the kinetic constants can then be determined from the intercepts on the axis, since when

$$\frac{1}{s} = 0,\ \frac{1}{v} = \frac{1}{V},$$

and when

$$\frac{1}{v} = 0,\ \frac{1}{s} = \frac{1}{K_m},$$

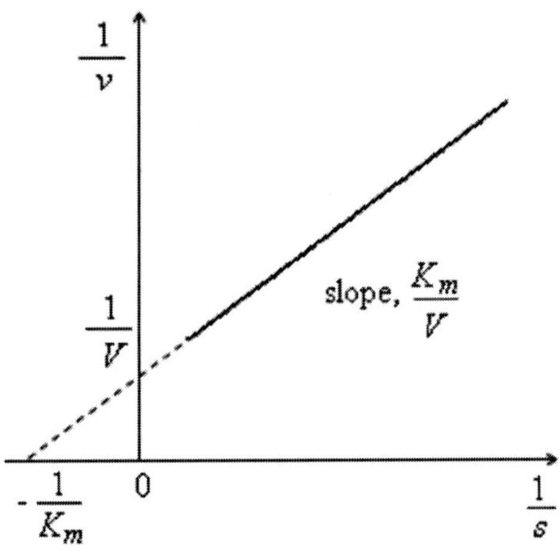

The determination of the Michaelis constant

An alternative plot is of s/v against s. This is obtained by multiplying the reciprocal equation by s:

when $$\frac{s}{v} = \frac{s}{V} + \frac{K_m}{V},$$

$$s = 0,\ \frac{s}{v} = \frac{K_m}{V}$$

and $$\frac{s}{v} = 0,\ s = -K_m,$$

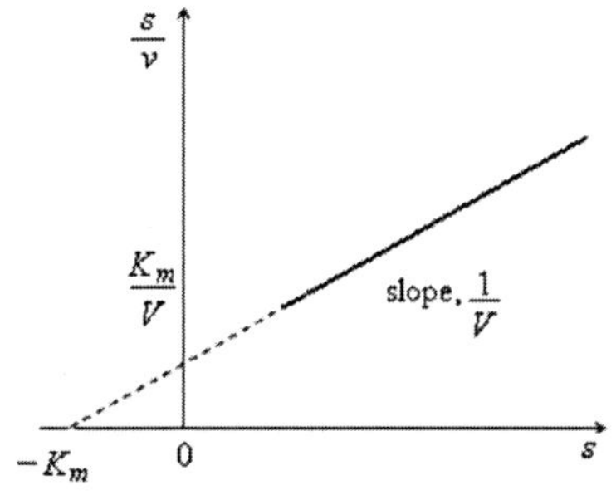

The determination of the Michaelis constant

The graph of (1/*v*) against (1/*s*) is known as the Lineweaver-Burk plot and is the method most frequently used to calculate K_m. However, the slope of the line is most influenced by the activities measured at low substrate concentrations, which are the least accurate, and, for this reason, the plot of (*s/v*) against *s* is to be preferred.

***Validity of the Michaelis equation*:** Most enzymes give a rectangular hyperbola when *v* is plotted against *s* in support of the Michaelis-Menten theory. There are some exceptions to this and these are dealt with in the next section on allosteric enzymes. There are, however, a number of important assumptions made in deriving the Michaelis equation and these may not always be valid.

These assumptions are as follows:

1. A complex is formed between the enzyme and the substrate and this break down to form the product.
2. Only one legend, namely the substrate, binds to the enzyme. Coenzymes, inhibitors and activators are ignored.
3. A steady state is rapidly attained and exists for the duration of the measurement.
4. The amount of substrate bound is small so that the concentration of free substrate is the same as the total present.
5. The total amount of enzyme present is the sum of the free enzyme and that bound in the ES complex.

Observations

....

....

....

....

Experiment – 4 : The Determination of the Michaelis Constant for the Digestion of Casein by Trypsin (3.4.4.4.)

Principle

The Michaelis constant of any of the enzymes whose assays are given in this book can be determined by measuring the activity over a range of substrate concentrations, but trypsin, in particular, is very convenient as the assay is a titration and can be conveniently carried out with a large class.

Trypsin is an enzyme found in pancreatic juice which catalyses the hydrolysis of peptide bonds formed from the carbonyl groups of the basic amino acids L-arginine and L-lysine. As the digestion proceeds, free amino groups are formed, so that the reaction can be followed by estimating the increase in the number of free amino groups using the formal titration.

Formaldehyde reacts with amino groups with the formation of methylol compounds and the liberation of H^+. The solution becomes more acid, and the amount of alkali required to neutralize this increased acidity gives a measure of the amount of free amino groups formed. Formaldehyde also very effectively stops any further activity by reacting with the enzyme protein.

$$-NH_3^+ + H \cdot CHO \rightleftharpoons -NH \cdot CH_2OH + H^+$$

$$-NH \cdot CH_2OH + H \cdot CHO \rightleftharpoons -N(CH_2OH)_2$$

Requirements

Reagents

1. 6 l Casein solution (40 g/l, pH 8.5). (Dissolve 40 g of casein About 900 ml of water containing 20 ml of 1 mol/l sodium hydroxide. Shake the suspension continuously and gently warm until dissolved. Finally, adjust the pH to 8.5 with 1 mol/ l HCl and make up to 1 l with water.
2. 1 l Trypsin (40 g/l). Crude trypsin prepared from pancreas is suitable.
3. 2 l Formaldehyde solution (400 g/l w/v).
4. 200 ml Phenolphthalein (2.5 g/l in alcohol).
5. 4 l Sodium hydroxide (0.1 mol/l).

Glass/Plastic wares: beakers, test tubes, flasks

Equipments: Water bath weighing balance.

The quantities are calculated on the assumption that this is a class exercise with several pairs given one concentration of casein.

Procedure

It is strongly recommended that a pilot run is carried out by the technical staff before the class so that the most convenient activity of the trypsin for the time course suggested can be obtained and the enzyme diluted if necessary.

Measure 5 ml of formaldehyde into six small flasks, add one drop of phenolphthalein to each one and 0.1 mol/l sodium hydroxide until the mixture is faintly pink. The colour in all the flasks should be the same.

Measure 100 ml of the casein solution into a conical flask and place in a water bath at 37^0 C for 10 min. Pipette 10 min of the trypsin solution, also previously warmed to 37^0C, into the casein and thoroughly mix. Carefully note the time. Immediately remove 10 ml of the mixture and run it into one of the flasks containing formaldehyde. Add one drop of phenolphthalein solution for every milliliter of reaction mixture taken and titrate with 0.1 mol/l sodium hydroxide until a faint but permanent pink colour is obtained. Just before the end point is reached, add an extra drop of indicator for every milliliter of sodium hydroxide taken. Withdraw further 10 ml samples of the digest at 2, 4, 6, 8, and 10 min and treat in exactly the same way. The final colour at the end point must be the same in each case. Plot the increase in titre against time and determine the initial reaction rate.

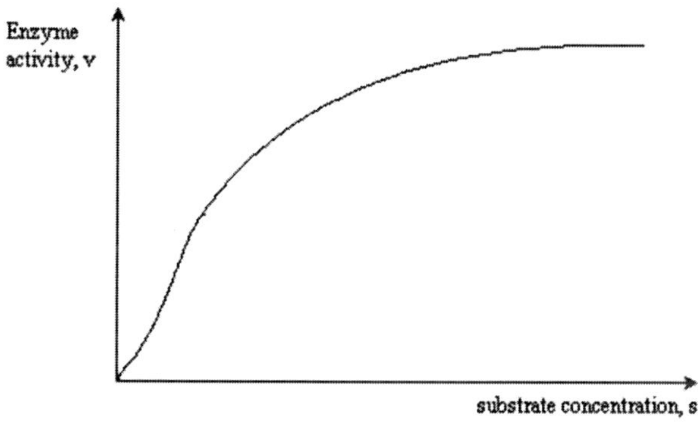

The effect of substrate concentration on the activity of an allosteric enzyme

Prepare casein solution of 5, 10, 15, 20 and 30 g/l and determine the activity at each of these substrate concentrations. Using your results, plot a graph of v against *s*, 1/v against 1/*s* and *s*/v against *s*. from your results; determine the values of V and K_m.

Allosteric Enzymes

Effects of substrate: There are some enzymes that give a sigmoid curve when the activity is plotted against substrate concentration and do not obey the usual Michaelis-Menten kinetics. These enzymes are made up of subunits and contain more than one active site per molecule. The sigmoid curve of *v* versus *s* can be explained on the basis that when one molecule binds, a conformational change takes place in the molecule enabling the next molecule of substrate molecules. Haemoglobin, although not an enzyme, is a good example of an allosteric protein with four binding sites for oxygen. The relative ease of binding of the oxygen atoms from the first to the fourth is approximately in the ratio of 1:4:24:9.

Metabolic control: Allosteric enzymes have specific sites for the binding of activators and inhibitors which act by causing a change in the conformation of the protein. This enables compounds that are chemically unrelated to the substrate to affect the enzyme activity, and allosteric enzymes are found at many points of metabolic control. For example, citrate is an allosteric inhibitor of phosphofructokinase, a key enzyme in the control of glycolysis, so that when citrate accumulates in the tricarboxylic acid cycle, the rate of glycolysis is reduced to avoid the build up of tricarboxylic acid cycle intermediates. Conversely AMP and ADP are allosteric activators of phosphofructokinase, so that when ATP concentration is low (ADP and AMP high) the rate of glycolysis is increased.

Allosteric kinetics: The derivation of an accurate equation for the sigmoid curve of allosteric enzymes is complicated, but a simplified equation can be obtained, as suggested by Atkinson, if the following assumptions are made:

1. The intermediates ES, ES_2, ES_3, etc., have only a transient existence.
2. Equilibrium is rapidly attained between the enzyme and the substrate molecules.

If there are *n* binding sites for the substrate S, then:

$$E + nS \underset{k_{-1}}{\overset{k_{+1}}{\rightleftharpoons}} ES_n \xrightarrow{k_{+2}} E + \text{Products,}$$

Assuming equilibrium,

$$K = \frac{k_{-1}}{k_{+1}} = \frac{[E][S]^n}{[ES_n]};$$

rearranging,

$$E = \frac{K\ ES_n}{S^n} \qquad (4\text{-}1)$$

Now if total amount of enzyme present is $[E_0]$, then:

$$[E_0] = [E] + [ES_n],$$

and, rearranging,

$$[E_0] = [E] - [ES_n],$$

Multiplying by k_3:

$$k_3[E] = k_3[E_0] - k_3[ES_n]. \qquad (4\text{-}2)$$

Now

$$v = k_3[ES_n] \qquad (4\text{-}3)$$

and

$$V = k_3[E_0] \qquad (4\text{-}4)$$

Substituting (4-3) and (4-4) in (4-2) we obtain:

$$k_3\ [E] = V - v. \qquad (4\text{-}5)$$

Substituting (4-1) in (4-5), then:

$$V - v = \frac{k_3K[ES_n]}{[S]^n} \qquad (4\text{-}6)$$

Substituting (4-3) in (4-6)

$$V - v = \frac{Kv}{[S]^n}$$

or

$$\frac{v}{V - v} = \frac{[S]^n}{K} \qquad (4\text{-}7)$$

Rearranging the equation,

$$v = \frac{V\ [S]^n}{K + [S]^n} \qquad (4\text{-}8)$$

If there is only one binding site for the substrate (n = 1) then this equation becomes the Michaelis equation and a hyperbolic plot is obtained.

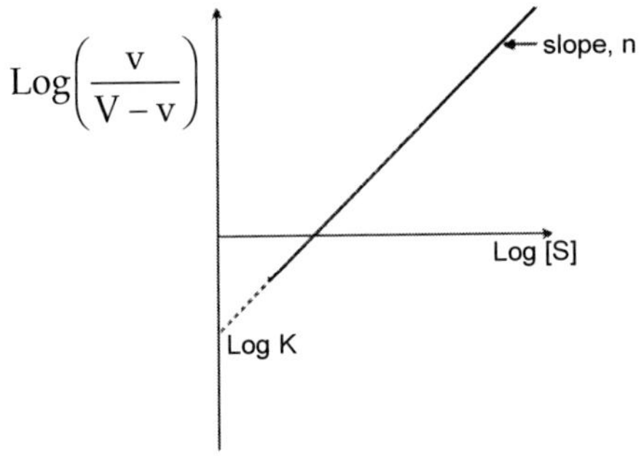

The determination of the kinetic constants for an allosteric enzyme

***Kinetic constants*:** A convenient linear transformation can be obtained by taking logarithms of Eq. (4.7). The kinetic constants can then be read off straight-line plot.

$\log v / (V - v) = n\log [S] - \log K$

n is a measure of the number of binding sites if the original assumptions are correct, but in practice n is usually less than this since the existence of the intermediates is not transient, n is the degree of cooperativity: the greater the value of n, the more sigmoid the plot of v versus [S].

K is actually a complex steady-state constant and for allosteric enzymes, when [S] = K, v is not $V/2$, but if v = $V/2$ then, from Eq. (4.7):

$$2\,[S]^n = K$$

The substrate concentration which gives 50% of the maximum activity is thus related to the constant K so that:

$$K = [S]^n_{50}$$

or, taking logs,

$$\log K = n \log [S]_{50}$$

Observations

....

....

....

....

Experiment – 5 : Yeast Isocitrate Dehydrogenase: An Allosteric Enzyme

Principle

Yeast isocitrate dehydrogenase catalyses the conversion of L-isocitrate to α-oxoglutarate and carbon dioxide with NAD as the coenzyme. It is a major control point in the tricarboxylic acid cycle and is modified by a number of effectors including NAD, AMP, and Mg^{++}. The enzyme is assayed by following the increase in extinction at 340 nm as NAD is reduced to $NADH_2$.

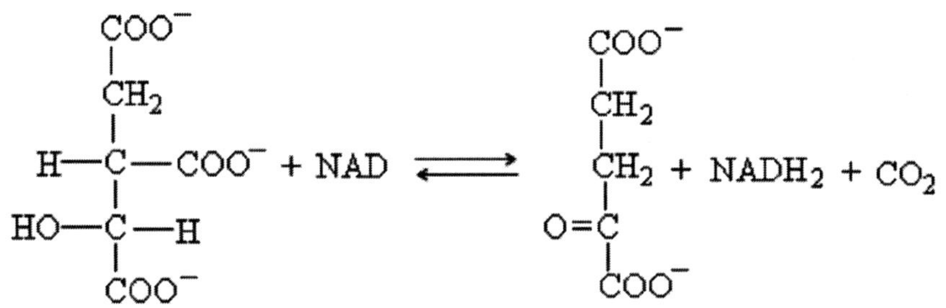

L-Isocitrate α-Oxoglutrate

*By permission of Dr P.J. Butterworth.

Requirements

Reagents

1.	Fresh bakers yeast	150 g
2.	Washed sand	-
3.	Sodium bicarbonate (0.1 mol/l)	200 ml
4.	Tris-HCl buffer (30 mmol/l, pH 7.4)	500 ml
5.	Sodium D, L-isocitrate in tris buffer, pH 7.4	200 ml
6.	NAD (2 mmol/l) in tris buffer, pH 7.4	250 ml
7.	AMP (3 mmol/l) in tris buffer, pH 7.4	50 ml
8.	$MgCl_2$ (0.1 mol/l)	50 ml

Glass/Plastic wares: Beakers, test tubes.

Equipments: Pestle and mortar, ultraviolet spectrophotometer, Circulating water bath, centrifuge, weighing balance.

Procedure

***Preparation of the enzyme*:** Prepare a crude extract of the enzyme by grinding some freshly grown yeast with washed sand and sodium bicarbonate solution (0.1 mol/l) in a pestle and mortar. The yeast cell wall is extremely tough and considerable grinding is required to break it. The proportions of yeast and sand are adjusted so that there is plenty of abrasive action by the sand, yet sufficient sodium bicarbonate present in which the released enzyme can dissolve. The grinding requires a little practice and it may be advisable for a technician to prepare a fresh extract for the class.

After grinding, transfer the mixture to centrifuge tubes, and remove the sand and large cell debris by centrifuging at 1000 g for 10 min. Collect the supernatant and recentrifuge at 30000 g for 1 h. Discard the precipitate and store the supernatant on ice until required.

***Enzyme assay*:** Prepare a series of tubes containing the following reaction mixture and incubate at 37^{0}C for 10 min.

NAD (2 mmol/l)	1.5 ml
$MgCl_2$ (0.1 mol/l)	0.1 ml
Diluted enzyme	0.2 ml

The amount of enzyme required has to be determined by experiment and if the preparation is too active it should be diluted with tris buffer.

Start the reaction by adding 1.2 ml of substrate previously incubated at 37^{0}C and follows the increase in extinction at 340 nm. Set up the appropriate blanks and controls to allow to any non-enzymic changes.

***Experiment*:** Measure the enzyme activity over a range of substrate concentrations from 0.1-1.2 mmol/l. This is achieved by triggering the reaction with from 0.1 to 1.2 ml of the substrate and adding tris buffer to give a final added volume of 1.2 ml. The substrate used is a mixture of the D and L forms so that the substrate used is a mixture of the D and L forms so that the L-isocitrate concentration ranges from 0.05 to 0.6 mmol/l.

Repeat the experiment but this time include AMP in the reduced per minute and prepare graphs of v against [S], 1/v against 1/[S] and log v/(V – v) against log [S].

If time permits, repeat the above experiment after heating the enzyme at 60^{0}C for 5 min. Another possible investigation is to examine the effect of Mg^{++} concentration on the reaction.

Coenzymes and Activators

In many cases, if an enzyme is mixed with its substrate under the appropriate conditions, either no catalysis occurs or there is only a slight activity. This is often due to the absence of a coenzyme or activator.

Coenzymes

These are low molecular weight organic compounds which are actively involved in the catalysis. They often act as acceptors or donors of specific chemical groups. NAD, for example, accepts and donates hydrogen atoms and is a coenzyme for many dehydrogenases. Coenzyme is the name given to soluble cofactors while the term prosthetic group is reserved for coenzymes that are firmly attached to the protein.

Activators

These are of a simple chemical nature and are not as specific as coenzymes. They appear to function by activating the enzyme substrate complex. Several metal ions are known to be activators of a wide range of enzymes; eg., Mg^{++}.

Observations

....

....

....

....

Experiment – 6 : Nicotinamide Adenine Dinucleotide as the Coenzyme for Lactate Dehydrogenase (L-lactate: NAD Oxidoreductase, 1.1.1.27)

Principle

Alkaline phosphatase is activated by divalent ions including Mg^{++}, Mn^{++}, Zn^{++}, Co^{++}, but especially Mg^{++}. The concentration of Mg^{++} for maximum activation is rather high at about 5 mmol/l and, because of this, it has been suggested that the magnesium salt of the phosphate ester is the true substrate.

Requirements

The material except that the calf intestinal enzyme is used instead of serum and $MgCl_2$ is also required.

Dilute the calf enzyme until a suitable activity is obtained with 10 mmol/l $MgCl_2$ present, then prepare a series of tubes with a range of $MgCl_2$ concentrations from 0 to 10 mmol/l and determine the activity. Prepare a graph of activity against $MgCl_2$ concentration.

Enzyme Inhibition

Types of Inhibition

Many compounds react with enzymes and reduce the measured activity. This property of enzyme is used in designing drugs and insecticides which selectively inhibit enzymes in the infective bacteria or insects, but do not affect the animal or plant. Two classical types of inhibition are recognized: competitive and non-competitive.

Competitive inhibition: In this case, the inhibitor reacts with the enzyme by competing with the substrate for the active site. The degree of inhibition depends on the relative concentrations of substrate and inhibitor, and almost maximal velocity may be found in the presence of the inhibitor if the substrate concentration is high enough.

Occasionally, the inhibition is found to be irreversible and the substrate cannot displace the bound inhibitor. This is the case with some of the organophosphorus inhibitors of cholinesterase. Competitive inhibition is also observed when the inhibitor binds at a site sufficiently close to the active centre to reduce the enzyme substrate affinity.

Competitive inhibitors are of a similar chemical structure to the natural substrate and are fairly. This is well illustrated in the case of the enzyme succinate dehydrogenase, which catalyses the conversion of succinate to fumarate. Malonate and maleate both act as competitive inhibitors of this enzyme.

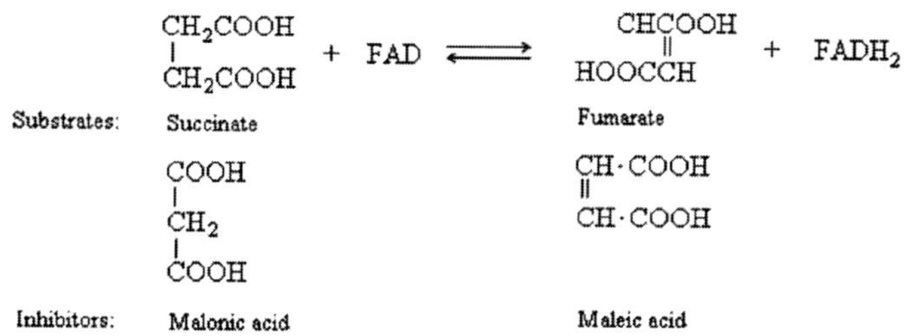

***Non-competitive inhibition*:** In the non-competitive type of inhibition, the inhibitor combines with the enzyme but not at the active site, so that the enzyme can bind both substrate and inhibitor at the same time. The binding site of the inhibitor is usually sufficiently far removed from the active centre so that the binding of the substrate is unaffected. The enzyme-substrate-inhibitor complex formed is unable to break down and inhibition effectively occurs by the reduction of the amount of enzyme available. Increase of the substrate concentration has no effect on the degree of inhibition.

Most non-competitive inhibitors are not related chemically to the substrate and the same inhibitor may affect a number of enzymes.

Examples of non-competitive inhibition are the action of thiol blocking agents such as *p*-chloromercuribenzoate, heavy metal ions such as Ag^{++} and Cu^{++}, and the reaction of cyanide with the iron-porphyrin enzymes.

***Determination of inhibitor constant* (K_i):** In the presence of an inhibitor, the Michaelis-Menten equation is altered as shown below, where *i* is the concentration of the inhibitor and K_i the inhibition constant:

No inhibitor, $$v = \frac{V}{1+(K_m/s)}$$

Competitive inhibitor, $$v = \frac{V}{1+K_m/s\,(1 + i/K_i)}$$

Non-competitive inhibitor, $$v = \frac{V}{(1+K_m/s)(1 + i/K_i)}$$

Taking reciprocals of the above equations we obtain:

No inhibitor, $\frac{1}{v} - \frac{1}{V} + \frac{K_m}{V}\frac{1}{s}$

Competitive inhibitor, $\frac{1}{v} - \frac{1}{V} + \frac{K_m}{V}\left(1 + \frac{i}{K_i}\right)\frac{1}{s}$

Non-competitive inhibitor, $\frac{1}{v} = \frac{1}{V}\left(1 + \frac{i}{K_i}\right) + \frac{K_m}{V}\left(1 + \frac{i}{K_i}\right)\frac{1}{s}$

A graph of (1/v) against (1/*s*) can be used to determine K_i and the type of inhibition. A competitive inhibitor alters K_m while the non-competitive type causes a change in V.

$$K_x = K_m\left(1 + \frac{i}{K_i}\right)$$

$$V_x = \frac{V}{(1 + i/K_i)}$$

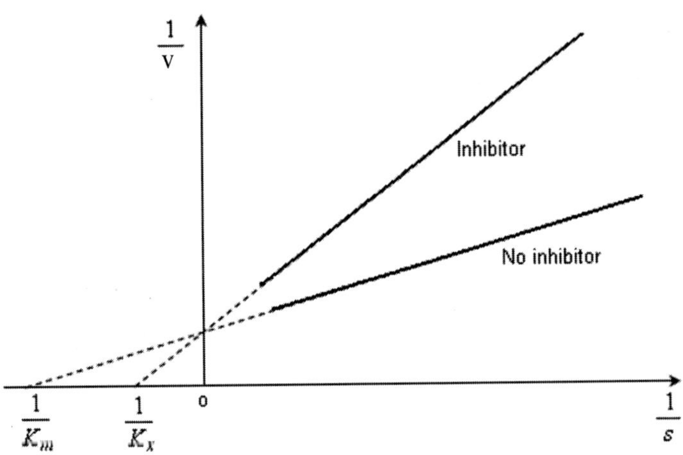

The effect of a competitive inhibitor on the Lineweaver-Burk plot

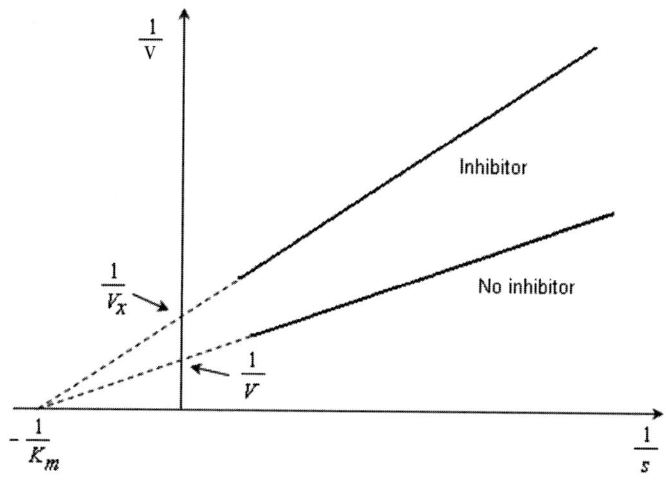

The effect of a non-competitive inhibitor on the Lineweaver-Burk plot

V and K_m are determined in the absence of an inhibitor and i is known, therefore K_i can be readily calculated.

Another useful way of determining K_i and the type of inhibition is to measure the activity of the enzyme at constant substrate concentration but with varying concentrations of the inhibitor. A graph is them plotted of (v/v_i) against i, where v is the activity with no inhibitor and v_i the activity in the presence of inhibitor of concentration i. The experiment is repeated at different substrate concentrations and the results compared.

For a non-competitive inhibitor: $\frac{v}{v_i} = 1 + \frac{i}{K_i}$

For a competitive inhibitor: $\frac{v}{v_i} = 1 + \frac{K_m}{s + K_m} \times \frac{i}{K_i}$

As can be clearly seen from the above equations, the graph of (v/v_i) against i for the non-competitive case is independent of substrate concentrations s, whereas the slope for a competitive type of inhibitor is dependent on s. This method is particularly useful if one has two substrates and wishes to know if they are affected by the same enzyme. The introduction of a non-competitive inhibitor will settle this question since, with a plot of ($v/v\text{-}_i$) against i, the points will be on the same line when both substrates are catalysed by the same enzyme.

In practice, other types of inhibition are often found apart from the simple cases considered.

Observations

....

....

....

....

Experiment – 7: The Effect of Inhibitors on Ox Heart Lactate Dehydrogenase

Principle

Lactate dehydrogenase catalyses the reversible reduction of pyruvate to lactate with $NADH_2$ as the coenzyme. The reduced coenzyme ($NADH_2$) absorbs strongly at 340 nm while the oxidized form(NAD) does not, so the progress of the reaction can be followed by measuring the decrease in extinction at 340 nm with pyruvate as substrate. The theory of this method is discussed more fully.

In this experiment, two compounds with similar structures to the natural substrate pyruvate are examined to see if they are competitive or non-competitive inhibitors

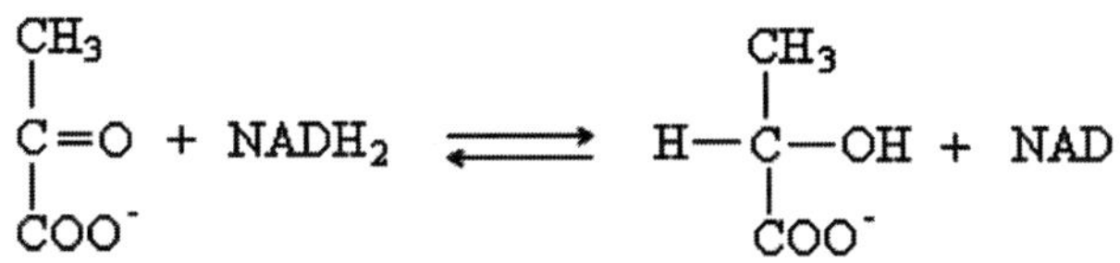

Substrates: Pyruvate L-Lactate

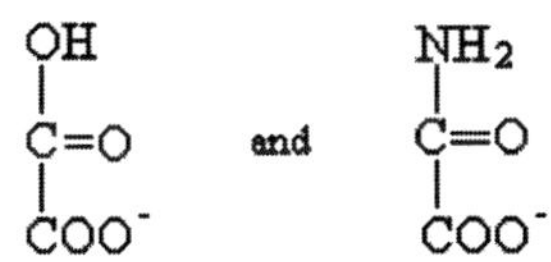

Inhibitors: Oxalate Oxamate

Requirements

Reagents

1. Phosphate buffer (0.1 mol/l, pH 7.4). 1l
2. Sodium pyruvate (21 mmol/l, prepared fresh in phosphate buffer). 200 ml
3. $NADH_2$ (3.5 mmol/l, prepared fresh). 25 ml
4. Ox heart lactate dehydrogenase diluted in phosphate buffer when required. 100 ml
5. Sodium oxalate (3 mmol/l, in phosphate buffer). 50 ml
6. Sodium oxamate (15 mmol/l, in phosphate buffer). 50 ml

Glass/Plastic wares: Beakers, test tubes.

Equipments: ultraviolet spectrophotometer, Circulating water bath, centrifuge, weighing balance.

Procedure

Initial assay: Prepare the following mixture in duplicate and equilibrate at 37^0C for 10 min.

Component	ml
Phosphate buffer	2.5
$NADH_2$ (3.5 mmol/l)	0.1
Ox Heart lactate dehydrogenase (diluted)	0.3

Rapidly add 0.1 ml of the sodium pyruvate solution also at 37^0C, transfer to a cuvette in a thermostatically-heated cell housing of an ultraviolet spectrophotometer, and observe the change in extinction at 340 nm with time. Dilute the enzyme preparation so that an extinction change in the region of 0.05-0.10 per minute is obtained.

Calculation of the activity: Enzyme activities are expressed most conveniently in nanomoles per minute per milliliter of enzyme or nanomoles per minute per milligram of protein. The molar extinction coefficient of $NADH_2$ at 340 nm is 6.22×10^3 ml mol^{-1} cm^{-1} so that a solution of 1 μmol/ml has absorption of 6.22.

For the 3 ml reaction mixture used in the assay,

Enzyme activity (μmol/min)

= Extinction change per minute/6.22 x 3.

Since 0.3 ml of enzyme solution is used in the assay,

Enzyme activity (μmol min^{-1} ml^{-1})

= Extinction per minute/6.22 x 3 x 1.0/0.3

= Extinction per minute x 1.61.

The final activity is obtained by multiplying by any dilution factor for micromoles per minute per milliliter and dividing by the protein concentration in milligram per milliliter for micromoles per minute per milligram.

Determination of the Michaelis constant: Dilute the sodium pyruvate solution ten times and repeat the above assay using volumes from 0.1 to 1.0 ml of the substrate to start the reaction. Adjust the initial volume of buffer in each case so as to give a final reaction mixture of 3 ml. Calculate the activity and plot graphs of v against s, 1/v against and s/v against and determine the K_m and V.

***The effect of inhibitors*:** (i) Repeat the above experiment but this time incorporate 0.1 ml of the inhibitor into the reaction mixture, again adjusting the volume of buffer to give a final reaction volume of 3 ml. Plot a graph of 1/v against 1/s and determine the inhibitor constant K_i. (ii) Repeat the initial assay but this time determine the activity at two fixed substrate concentrations and a range of concentrations of the inhibitor, up to a final concentration of 0.1 mmol/l oxalate and 0.5 mmol/l oxalate. If v_i is the activity in the presence of the inhibitor and v is the activity in the absence of the inhibitor, plot a graph of v/v_i against i. Use these graphs to decide what type of inhibition is given by oxalate and oxamate and determine the inhibitor constants.

***High substrate concentration*:** Examine the effect of high substrate concentrations on a plot of v against s and s/v against s. This is carried out by using volumes up to 1 ml of 21 mmol/l sodium pyruvate to trigger the enzyme reaction. Is there anything unexpected about the graphs? If so try to explain what is observed.

Temperature and pH

The Effect of Temperature on Enzymes

The rate of enzyme-catalysed reaction, like most chemical reactions, increases with temperature. This is due to the effect on the velocity constants of the various parts of the overall reaction such as k_{+1}, k_{-1}, k_{+2}, and the affinity of the enzyme for cofactors, activators etc. The pK values of ionizable groups concerned in the reaction are also affected by temperature, but this is not easy to interpret and will not be considered in any further detail. At elevated temperatures the enzyme protein is denatured, resulting in a loss of activity. This means that the initial reaction rate will rise with temperature until the becomes impossible to measure due to almost immediate inactivation. In practice, most enzymes are completely inactivated above 70^0C.

If, however, the enzyme 'activity' is incorrectly measured by determining the amount of substrate transformed in a given time at different temperatures, then a so-called 'optimum temperature' is obtained.

The 'optimum temperature' is the result of the balance between the rate of increase of 'activity' and the rate of destruction of the enzyme. The 'optimum temperature' is not a constant value for a given enzyme, but depends upon the time during which measurements of activity are made. The shorter the time of measurement, the higher will be the apparent optimum temperature.

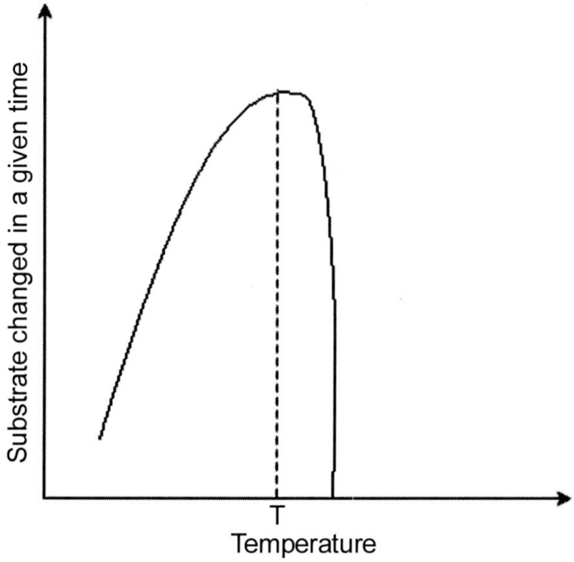

The 'optimum temperature' (T) of an enzyme

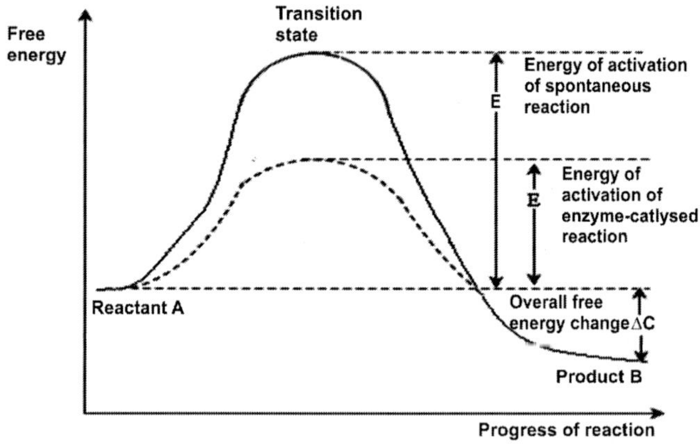

Energy diagram of the reaction A → B in the presence of (--------) and absence (--------------) of an enzyme

The effect of temperature on the enzyme reaction: Molecules must possess a certain energy of activation (E′) before they can react, and enzymes function as catalysts by lowering this energy of activation (E), thereby enabling the reaction to proceed more rapidly. The overall change in the *free energy* (ΔG) is, of course, unaffected.

The energy of activation can be determined by measuring the maxima velocity at different temperatures, and plotting $\log_{10}$ V against 1/T. The slope of the line is given by E/2.303 R. This relationship is obtained from the empirical equation of Arrhenius:

$$d\ln k/dT = E/RT^2 \cdot$$

Integrating this equation one obtains:

$$\log_{10} k = C - E/2.303RT,$$

where C = a constant, k = velocity constant of the reaction, T = absolute temperature, R = gas constant (8.32×10^{-3} J mol^{-1} K^{-1}), E = energy of activation (J/mol).

The reaction constant is not always easy to obtain so, as the maximum velocity is directly proportional to *k*, this activity is usually plotted against the reciprocal of the temperature.

The Arrhenius equation is similar to that derived from the theory of absolute reaction rates:

$$d\ln k/dT = (\Delta H + RT)/RT^2 \cdot$$

From these equations, it is clear that the energy of activation E = ΔH + *RT*, where ΔH is the heat of activation or enthalypy of the reaction.

***The effect of temperature on the enzyme protein*:** This can readily be determined by exposing the enzyme to a high temperature and measuring the activity at a temperature where the enzyme is stable. This will tell us how much of the enzyme has been destroyed. A graph is prepared at each temperature of the change in active enzyme with time. The initial rate v is obtained from this graph, which is then repeated for a number of temperatures. The energy and hence the heat of inactivation can be obtained as previously described from a plot of $\log_{10} v$ against $1/T$. The heat of inactivation ΔH is usually quite high as a result of the large positive entropy change resulting from the opening up and unfolding of the protein molecule during denaturation.

Observations

....

....

....

....

Experiment – 8 : Determination of Nitrate Reductase Activity

Principle

The reduction of nitrate to nitrite is catalysed by the enzyme nitrate reductase, which uses reduced pyridine nucleotides as electron donor (coenzyme).

$$NO_3^- + 2e^- \underset{NAD(P)H + H^+}{\overset{\text{Nitrate reductase}}{\rightleftharpoons}} NO_2^- + H_2O$$

The enzyme nitrate reductase was first studied by A. Nason and H. J. Evans (1953) in the mold, *Neurospora crassa.* Since then, it has been demonstrated in almost all plants examined. It has been isolated and purified from higher plants as well. The enzyme is a homodimer with monomer size 100 to 120 k Dalton. Although it may very considerably among different species. Molybdenum is an essential constituent of the enzyme and flavin adenine dinucleotide (FAD) is presented as a lightly bound prosthetic group of the enzymic protein. Heme in the form of cytochrome b-557 also seems to be an integral component of higher plant enzyme.

Requirements

Reagents: distilled water, phosphate buffer, KNO_3, sulphanilamide, napthyl ethylene diamine, dihydrochloride, EDTA, caseine, NADH, Fresh sample (leaves) tiny pieces.

Glass/Plastic wares: Beakers, test tubes, funnel, measuring cylinder, petridish, conical flask,

Equipments: Pestle and mortar, ultraviolet spectrophotometer, Circulating water bath, centrifuge, weighing balance, filter paper,

Procedure

Nitrate reductase activity measure by either *in vivo* or *in vitro* procedure.

***In vivo* enzyme assay:** One g fresh sample from roots, leaves or cotyledons take and excise into tiny pieces (1 – 2 mm). The cut segments incubate in a assay mixture containing 8.0 ml of 0.1M phosphate buffer of 7.4 pH, 1.0 ml of 0.2 M KNO_3 and 1.0 ml of 25% (v/v) propanol in a black taped air tight vials. The incubation mixture is first evacuate and then the vials are incubate in dark at 30^0C for 30 min. the nitrite produces thereafter estimate colorimeterically. For this to 2.0 ml of incubation medium, 2.0 ml napthyl ethylene diamine dihydrochloride (0.2% w/v in distilled water) add. The intensity of pink colour (a diazo complex) so obtain is measure after 15 min at 540 nm by a spectrophotometer. The amount of nitrite is calculate using a standard curve of KNO_2 (Fig. 8.1).

***In vitro* enzyme assay:** 500 mg fresh tissue is extract with 2.0 ml of cooled (0 – 4^0C) extraction medium which consist of 2.0 M phosphate buffer of 7.5 pH, EDTA 0.2 M, 1 mM of cysteine and 0.5% caseine. The extract is then centrifuge at 12,000 x g for 20 min. the clear supernatant is used as enzyme preparation. The assay mixture consist of 2.0 ml of 0.2 M sodium phosphate buffer (pH 7.8), 0.4 ml of 0.2 M KNO_3, 0.3 ml of freshly prepared (in buffer) 2 mM NADH and 0.4 ml of the enzyme preparation. The incubation carries on for 30 min at 30^0C and the reaction is terminate by adding 1.0 ml of 1.5% sulfamilamide in 1.5 N HCl. One ml of napthylene ethylene diamine dihydrochloride add immediately thereafter and the absorbance of the pink colour mixture is record at 540 nm and the amount of nitrite form is calculate with the help of the standard curve of KNO_2 (Fig.8.1).

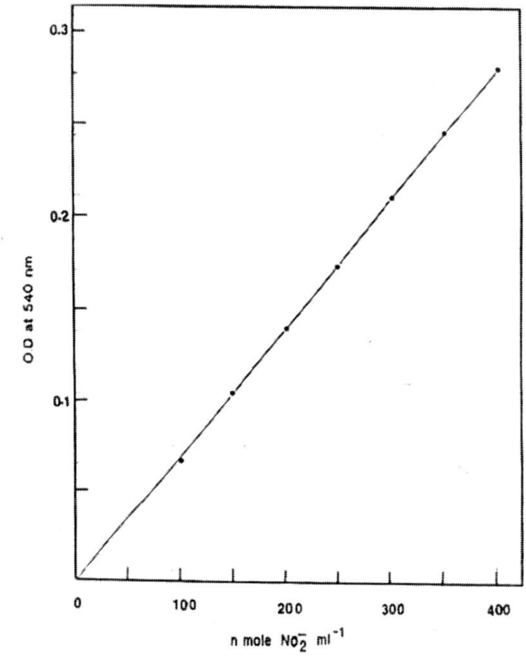

Fig. 8.1 : Calibration curve for nitrite from potassium nitrite solution.

Observations

....

....

....

....

Experiment – 9 : Demonstrate the Diastase Activity in Germinating Seeds of Barley

Principle

Diastase is the enzyme which acts on starch and convert it to a hexose as portrayed by the reaction below.

$$2n\ (C_6H_{10}\ O_2{+}nH_5)\ O \longrightarrow nC_{12}H_{22}0_{11}$$
$$C_{12}H_{22}O_{11} + H_2O \longrightarrow 2\ C_6H_{12}O_6$$

Requirements

Reagents

Starch solution. Take 5 ml of starch solution (2 gm of soluble starch in 50 ml boiling water) in a flask of 200 ml capacity.

Buffer solution A. It is prepared by taking 6.95 gm of monobasic sodium phosphate (0.2 M) in 250 ml of distilled water.

Buffer solution B. 13.41 gm of $Na_2HOP_4.7N_2O$ or 17.92 mg of $Na_2HPO_4.12N_2O$ (dibasic sodium phosphate) is dissolved in 250 ml of water to get buffer solution B (0.2 M). 1% iodine solution in KI(iodine=1 gm, KI=2 gm, and water =300 ml).

Pipette 39 ml of stock buffer A and 61 ml of stock buffer B to this and make up the volume. This will give 1% starch solution in neutral buffer at pH 7.0.

Glass/Plastic wares: beakers, test tubes, measuring flask, microtips

Equipments: mortar and pestle, micropipette, water bath, cheese cloth, centrifuge

Procedure

Grind 10 gm of germination barley seeds in twice as much of water in mortar. Filter through cheese cloth (or muslin) and centrifuge the filtrate at low speed. Take the supernatant liquid in a measuring flask and make up the volume of the enzyme extract to 100 ml.

Take 1 ml of 1% iodine solution prepared in potassium iodine in six test tubes and add 20 ml of water to each of them. For a control add 1 ml starch solution in a test tube containing 1 ml of 1% iodine solution (in KI) and 20 ml of water. Measure 10 ml of starch in a beaker add 1 ml of enzyme extract and shake well. Pipette out 1 ml of this starch diastase mixture (now known as digestion mixture) into each of the six test tubes containing the iodine solution.

Note the time of the disappearance of the colour. After 10 min of digestion pipette out 1 ml of digestion mixture into a test tube and test for sugar with Benedict's solution.

Effect of Variables like pH, Temperature or Concentration of Diastase on Its Activity

pH : Take 5 ml of buffer solution of known pH(5.0,5.5,6.0,6.5,7.0,7.5, 8.0, 8.5, 9.0) in separate test tubes. Add 5 ml of 1% starch solution in each test tube. Add and note the time of addition of 1 ml of diastase solution to the test tubes. Mix the contents of the tube well. Pipette 1 ml every 5 min of the reaction mixture into separate tubes containing 1 ml of iodine solution (in KI) and 20 ml of water. Plot the time taken (in min) for complete hydrolysis as shown by disappearance of the iodine colour) against the pH on a graph paper.

Temperature : Maintain seven water baths at 0, 10, 20,40, 60,80, and 100C. Take 7 test tubes and add in each 5 ml of 1% soluble starch (ph 7.0) and place one each in baths of the different temperatures. When the tubes and their contents have attained the temperature of the bath note down the time and add 1 ml of diastase solution in every test tube. Shake well. After an interval of 5 min add 1 ml this mixture in test tubes containing 1 ml iodine solution +20 ml of water. Plot the time (in min) taken for disappearance of the colour of iodine against temperature. Work out Q_{10} of diastase (rise in enzyme activity for every rise of 1°C) does Vant Hoff's Law hold good? If so, state the range of temperature for which it holds good.

Concentration of enzyme : Crush in mortar 10 gm germinating barley seeds with 20 ml of water. Filter through muslin cloth and centrifuge the filtrate. Decant the supernatant liquid and make up the volume to 100 ml. This is the source of amylase for this experiment.

Take 10 empty test tubes and to each add 5 ml of starch and 5 ml of buffer solution (pH7). Add 1 ml of 13 dilutions of amylase solution: H_2O : : 2:0, 2:2, 2:4, 2:6, 2:8, 2:10, 2:12, 2:14, 2:16, 2:18, 2:20, 2:25, 2:30). Shake well, record the time, and immediately pipette 1 ml of starch, enzyme mixture into tubes containing 1% iodine solution (in KI) +20 ml of water every 5 min. Note the time of disappearance of colour. Plot the results with concentrations on the horizontal axis and time on vertical axis.

Observations

....

....

....

....

Experiment – 10 : The Determination of the 'Optimum Temperature' of α-amylase

Principle

α-Amylase catalyses the hydrolysis of α 1-4 links of starch with the production of reducing sugars. The reaction is followed by measuring the increase in reducing sugars using the 3,5-dinitrosalicylate reagent when an alkaline solution of 3,5-dinitrosalicylic acid is reduced to 3-amino-5-nitrosalicylic acid. The reaction is followed by measuring the extinction at 540 nm.

This exercise and the next one are best carried out as class projects because of the large number of measurements to be taken. The quantities required are calculated on the assumption that each pair will assay in duplicate at one temperature only.

Requirements

Reagents

1. Sodium or potassium phosphate buffer (0.1 mol/l, pH 6.7). 5l
2. Buffered starch substrate (5g/l in phosphate buffer. Mix 5g of soluble starch to a smooth paste with about 50 ml of boiling phosphate buffer solution to boil for 1 min, then cool to room temperature, and dilute to 1 l with buffers solution. 3l
3. Sodium chloride (10 g/l). 3l
4. α-Amylase suitably diluted. 1l
5. Sodium hydroxide (2 mol/l). 1l
6. Sodium potassium tartrate (Dissolve 300 g of this salt in about salt in about 500 ml of water). 500 ml
7. 3,5-Dinitrosalicylic acid (Dissolve 10 g of this reagent in 200 ml of 2 mol/l sodium hydroxide). 200 ml
8. Dinitrosalicylate reagent (Prepare this by mixing solutions 8 and 9 and making up to 1l with water). 1l

Glass/Plastic wares: beakers, test tubes, microtips, measuring cylinder.

Equipments: Colorimeter, Water baths, micropipette.

Procedure

Set up eight tubes containing the following reaction mixture:

Contents	ml
Starch (5 g/l)	2.5
Phosphate buffer (0.1 mol/l, pH 6.7)	1.0
Sodium chloride (10 g/l)	0.5

Place the tubes in a water bath at a fixed temperature and equilibrate the enzyme at the same temperature. After 10 min add 0.5 ml of the enzyme to seven of the tubes and 0.5 ml of water to the blank. Incubate the tubes for 0, 5, 10, 15, 20, 30 and 40 min and stop the reaction by adding 0.5 ml of 2 mol/l NaOH to all tubes. Add 0.5 of the dinitrosalicylate reagent and heat the tubes for 5 min in a boiling water bath, cool, and read the extinction at 540 nm against a blank. *Note*: The tubes must all be cooled to room temperature before reading since the extinction is sensitive to temperature change.

Plot the extinction of each solution against the time of incubation and prepare a progress curve of the reaction. Collect the data from the other groups and plot a graph of the change in the initial reaction rate against temperature and also the substrate changed per minute at 10 and 40 min against temperature. Compare the graphs and explain them.

Plot a graph of $\log_{10} v$ against $1/T$ and determine the heat of activation. What assumptions have been made in this experiment?

For convenience the activity can be expressed in terms of extinction change at 540 nm per minute.

Observations

....

....

....

....

Experiment – 11 : The Effect of Temperature on the Stability of α-amylase

Principle

Starch is composed of two components, namely amylose (15-20%) a long unbranched straight chain component and amylopectin (80-85%) a branched chain polysaccharide. Starch from waxy corn is notable as it consists practically of amylopectin component, there being no amylase. α- amylase or simple amylase has a molecular weight range of 10,000 to 50,000. It may be formed in plants cells by elimination of a molecule of water from glycosidic OH group of one α-D glucose molecule and alcoholic OH group on carbon 4 of the adjacent α-D glucose molecule. The linkage in amylase is thus an α-1,4-glucoside, like that in maltose. Enzymic hydrolysis of amylase with amylase, henceforth, yields maltose units mainly. Amylase may be considered as an anhydride of α-D-glucose units.

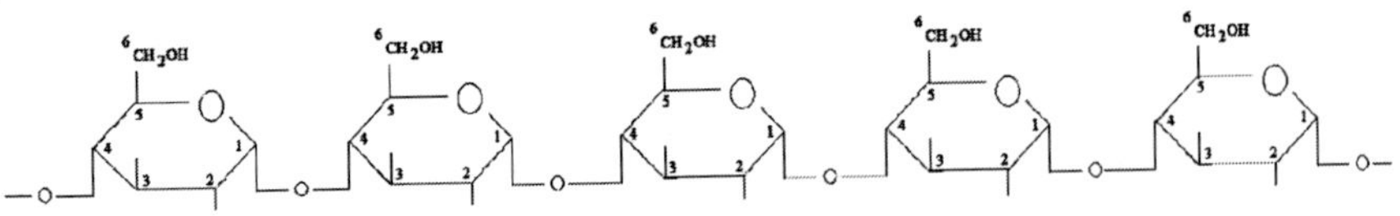

Structure of Amylose (= A - form)

Various types of structures differ in the amount of the amylase component present in them (table-11.1).

Table 11.1 : Amylose contents of some starches.

Starch	Amylase component %
Sugary mutant corn	70
Steadfast pea	67
Alderunan pea	65
Buckwheat	28
Barley	27
Sorghum	27
Commercial corn	26
Wheat	25
White potato	23
Arrowroot	21
Sweet potato	20
Tapioca	18
Waxy barley	3
Waxy corn	0-6

Requirements

The reagents and quantities are the same as in the previous experiment No 10, assuming that the exercise is carried out as a class project.

Procedure

Prepare a suitable dilution of the α-amylase so as to give a good activity when incubated 5 min at 37^0C. Place the enzyme in a boiling tube and equilibrate for 10 min. Pipette 0.5 ml of the enzyme into four test tubes in a water bath at a fixed temperature from 50 to 75^0C. Remove in the tubes after 5, 10, 15 and 20 min (or other suitable time intervals) and rapidly cool under running tap water. Incubate the tubes at 37^0C and assay for enzyme activity after the addition of the substrate as in the previous experiment. The mixture is incubated for 5 min and the activity determined with 3,5-dinitrosalicylate. The rate of inactivation of the enzyme can then be determined. Collect the data from the groups who have looked at other temperatures and prepare a graph of the rate of inactivation (v) against $1/T$ and determine the heat of inactivation. Compare this figure with that obtained for the heat of activation from the last experiment.

pH and Enzyme Activity

The pH optimum: Enzymes are active over a limited pH range only and a plot of activity against pH usually gives a bell-shaped curve of the type shown in figure. The pH value of maximum activity is known as the optimum pH and this is a characteristic of the enzyme, provided that the enzyme is stable under the conditions studied.

The variation of activity with pH is due to the change in the state of ionization of the enzyme protein and other components of the reaction mixture. Michaelis and Davidsohn suggested in 1911 that only of the large number of ionized forms of the protein is active, so that a change in pH either side of the optimum produces a decrease of this form and hence a fall in the activity.

K_m and V. Changing the pH alters the enzyme activity by affecting V, K_m, or the stability of the enzyme protein. Most published plots of pH optima are the result of changes in V and K_m and are not, therefore, readily amenable to mathematical treatment. Ideally the effect of pH on K_m and V should be studied separately.

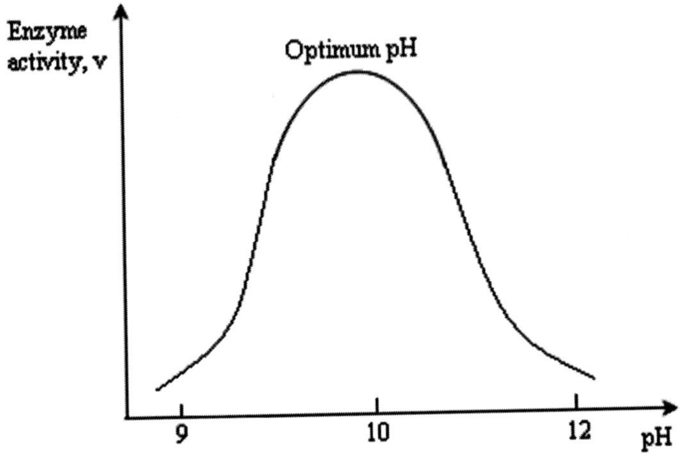

The pH optimum of an enzyme illustrated by rat alkaline phosphate

***Enzyme stability*:** If the enzyme is unstable at certain pH values, then the optimum pH is no longer a characteristic of the enzyme. The stability can be checked by exposing the enzyme to the appropriate pH value for the time of the experiment and then adjusting the pH to a value at which the enzyme is stable and measuring the activity.

Observations

....

....

....

....

Experiment – 12 : The pH Optimum of an Enzyme

Materials and Method

Determine the pH optimum of alkaline phosphatase or α-amylase using the materials and assay methods. For alkaline phosphatase use a range of 0.1 mol/l sodium carbonate-bicarbonate buffers from pH 9 to 11 and prepare a standard curve at each pH value. In the case of α-amylase, use sodium or potassium phosphate buffer solutions from pH 5.8 to 8.0.

Observations

....

....

....

....

Experiment – 13 : Determination of Enzyme Glutamate Dehydrogenase Activity (GDH) from Plant Leaves/Shoot/ Root

Principle

The enzyme is present in large quantities in plant tissues and uses either NADH or NADPH as coenzymes. The enzyme also catalyzes the reversible reaction i.e. hydrolysis of glutamate to 2-oxo glutarate and ammonia and uses either NAD or NADP as coenzymes.

$$\text{2-oxo-glutarate} + NH_4^+ + NAD(P)H^+ \underset{}{\overset{GDH}{\rightleftharpoons}} \text{Glutamate} + NAD(P) + H_2O$$

Various studies have shown that enzyme prefers NADH over NADPH and NAD over NADP by various degrees. The enzyme is thought to be a hexameric metallo-protein having molecular weight in the range of 208,000 to 270,000. The metallic component is zinc.

The role of glutamate dehydrogenase in the synthesis of glutamic acid has been doubted by various workers, because it catalyses the hydrolysis of glutamic acid as well and because its affinity for ammonia is very low. However it appears that the enzyme plays an active role in the assimilation of ammonia under stress conditions, such as dark starvation, high temperature, salinity, water stress etc.

Procedure

Extraction

Glutamate dehydrogenase (GDH) was extracted in a mortar in a medium containing 0.05 M sodium phosphate buffer (pH 7.4), 0.4 M sucrose, 2mM EDTA (as sodium salt). The ratio of the plant tissue to medium was 1.4 (w/v). The samples were thoroughly extracted at 0-4^{0}C filtered through two layers of cheese cloth and centrifuged at 6,000 rpm at the same temperature for 15 min. the clear supernatant was used as enzyme preparation.

Assay

The enzyme activity in the pellets was less than 10% of the supernatant; hence the activity in supernatant was routinely assayed spectrophotometrically for aminating activities of GDH.

Assay of NADH-Dependent Aminating Activity

The assay mixture of aminating enzyme (NADH-dependent) consisted of 4.0 ml of 0.1 m sodium phosphate buffer (pH 8.1), 0.6 ml of 0.2 m 2-oxoglutarate, 0.6 ml of M ammonium sulphate, 0.6 ml of 1.5 M ammonium sulphate, 0.6 ml of mM NADH and 0.1 ml of enzyme preparation giving a final volume of 6.0 ml. each component of the assay mixture was dissolved in the buffer. The reaction was started by adding the substrate ammonium sulphate and the decrease in absorbance was recorded continuously for 5 min at 340 nm in a Bechman DU64 spectrophotometer (Singh and Srivastava, 1982).

The amount of NADH oxidized was calculated from standard curve for NADH. The oxidation of NADH with enzyme preparation was linear upto 5 min atleast. Representative graphs of enzyme activity from root and shoot tissues have been given (Fig 13.1).

Partial Purification of the Enzyme

In some experiments, partially purified enzyme was used. The fresh plant materials were homogenized in a mortar. The crude homogenate was brought to 30% ammonium sulphate saturation for 20 min. at 0-4^{0}C and centrifuged at 6,000 rpm for 10 min. Ammonium sulphate was again added to the supernatant to produce 55% saturation and after keeping for 20 min in cold it was centrifuged as above. The precipitate formed between 30 and 55% saturation of ammonium sulphate was redissolved in 1/10 of its original volume in 0.05 M sodium phosphate buffer (pH 7.4). The sample was desalted on a sephadex G-100 column equilibrated with the same buffer. The protein eluted was collected in 2 ml fractions. The fractions containing maximum enzyme activity were pooled and used for the assay. A typical purification data from root and shoot sample. It is apparent that upto 3 to 4 folds purification was possible using the procedure described.

Preparation of Sephadex-Column

About 2.0 g of the dry Sephadex G-100 (Sigma Chemical Co., U.S.A.) was soaked at room temperature in 100 ml of distilled water for 3 days. After that the mixture was stirred with a glass rod for 2 min and then allowed to settle. The fine particles floating on the surface were decanted away. The process of washing and decantation was repeated 3 to 4 times.

The suspension of Sephadex was taken in a conical flask connected with a vaccum pump. The air present in the sephadex suspension was removed by evacuating the flask. The glass column used was 40 cm long with a diameter of 1.5 cm. A small amount of glass wool was placed at the bottom of the column and Sephadex suspension was packed slowly with the use of a pipette through the opposite end and with the help of a clamp adjusted to regulate the speed of the eluent. The column was equilibrated with 25 mM sodium phosphate buffer pH (7.5) before use.

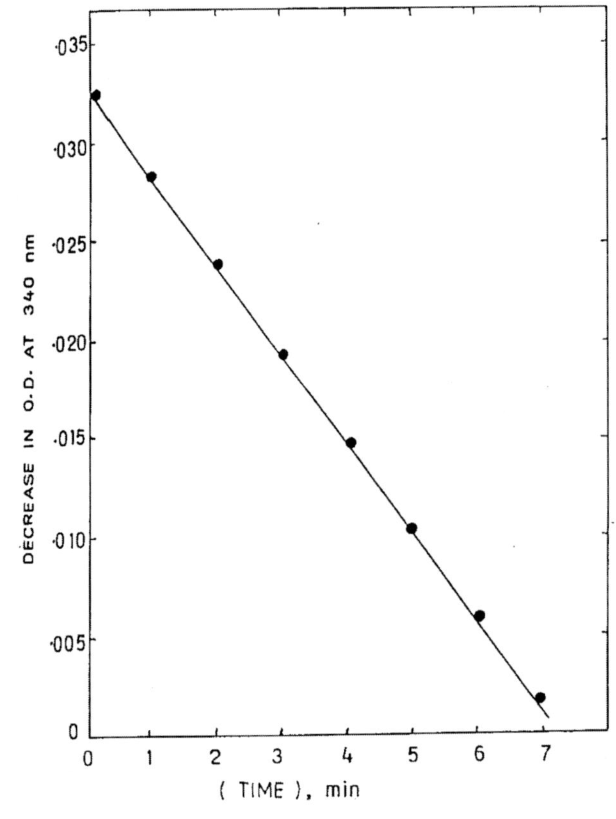

Fig.13.1 : Decrease in absorbance due to oxidation of reduced nicotinamide adenine dinucleotide (NADH), by glutamate dehydrogenase.

Observations

....

....

....

....

Experiment – 14 : Determination of Enzyme Glutamine Synthetase Activity (GS) from Plant Tissue

Principle

The ammonia is converted to glutamic acid by two consecutive reactions is through the synthesis of amide glutamine.

$$\text{(1) Glutamic acid} + NH_4^+ + \text{ATP} \underset{\text{Glutamine synthetase (GS)}}{\overset{Mg^{3+}}{\rightleftharpoons}} \text{Glutamine} + \text{ADP} + \text{Pi} + H^+$$

$$\text{(2) Glutamine} + \text{2 - oxoglutarate} + \text{NADPH/FD red} + H^+$$

$$\xrightarrow[\text{Glutamate synthase (GOGAT)}]{} \text{2 Glutamic Acid} + \text{NADP/FD}$$

The first reaction in the series is catalysed by the enzyme glutamine synthetase while the second one is catalysed by glutamate synthetase.

Synthesis of glutamine from glutamate and ammonia occurs in both roots and leaves. The enzyme glutamine from glutamate and ammonia occurs in both roots and leaves. The enzyme glutamine synthetase dehydrogenase. However amount of glutamate is always necessary to drive this reaction.

Procedure

Activity of glutamine synthetase was assayed in the crude extracts by the method of Lillo (1984) with slight modifications.

Preparation of Extracts

Extracts were prepared at 0-4^{0}C. Root and shoot tissue (1.0g each) was homogenized with sand in a mortar with 4 ml of 50 mM Tris HCl (pH 7.8) containing 15% (v/v) glycerol, 14 mM 2-mercaptoethanol, 1 mM EDTA and 0.1% (v/v) Triton X-100. The homogenate was squeezed through two layers of Mira-cloth and then centrifuged at 5000 r.p.m. for 10 min at 0^{0}C. The supernatant was used as the enzyme extract.

Assay of the Enzyme

Glutamine synthetase (GS) was determined using the synthetase assay (Webster, 1964). The standard reaction mixture contained 500 mM L-glutamate, 100 mM ATP, 300 mM $MgSO_4$, 200 mM NH_2OH and 200 mM Tris HCl in a total volume of 2.0 ml. The final pH was 8.0. 0.1 ml of enzyme extract was added to start the reaction giving a total assay volume of 2.0 ml. After incubation at 27^{0}C for 30 min the reaction was terminated by adding 4.0 ml 2.5% (w/v) $FeCl_3$ and 5% (w/v) trichloroacetic acid in 1.5 M HCl. After centrifugation the absorbance of the supernatant was read at 540 nM in Beckman DU-64 spectrophotometer. The enzyme activity was calculated using a standard curve prepared form Glutamyle hydroxamic acid (Fig. 14.1).

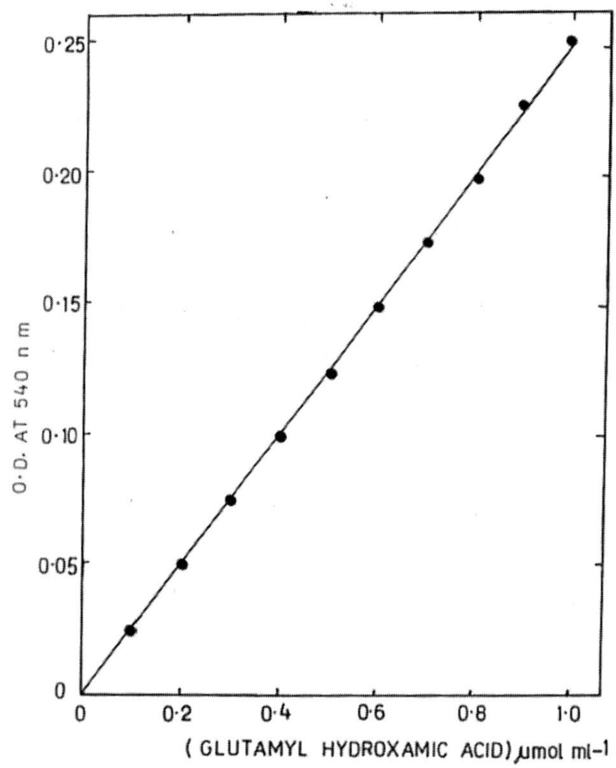

Fig. 14.1 Calibration curve for glutamyl hydroxamic acid (GHA).

Observations

....

....

....

....

Experiment – 15 : Determination of Enzyme Glutamate Synthase Activity (GOGAT) from Plant Tissue

Principle

The ammonia is converted to glutamic acid by two consecutive reactions is through the synthesis of amide glutamine.

$$\text{(1) Glutamic acid} + NH_4^+ + \text{ATP} \underset{\text{Glutamine synthetase (GS)}}{\overset{Mg^{3+}}{\rightleftharpoons}} \text{Glutamine} + \text{ADP} + \text{Pi} + H^+$$

$$\text{(2) Glutamine} + 2\text{ - oxoglutarate} + \text{NADPH/FD red} + H^+$$

$$\xrightarrow[\text{Glutamate synthase (GOGAT)}]{} \text{2 Glutamic Acid} + \text{NADP/FD}$$

The first reaction in the series is catalysed by the enzyme glutamine synthetase while the second one is catalysed by glutamate synthetase.

Synthesis of glutamine from glutamate and ammonia occurs in both roots and leaves. The enzyme glutamine from glutamate and ammonia occurs in both roots and leaves. The enzyme glutamine synthetase dehydrogenase. However amount of glutamate is always necessary to drive this reaction.

The second enzyme glutamate synthase catalyses the transfer of amide group of glutamine to 20x0 glutamate producing two molecules of glutamate. It uses either reduced ferredoxin or NAD(P) H as reductant. The glutamic acid arising from nitrate assimilation is converted into many other amino acids. It can also be incorporated into proteins.

Procedure

The enzyme was extracted and assayed in the crude extract by the method of Singh and Srivastava (1986).

Preparation of Extracts

Tissue was extracted in a mortar in a medium containing 0.2 M sodium phosphate buffer (pH 7.5), 2 mM EDTA, 50 mM KCl 0.1% mercaptoethanol and 0.5% triton X-100 wit an approximate amount of washed sand, in a ratio of 1:4 (w/v) at 0-4^{0}C and filtered through 2 layers of cheese cloth. The homogenate was centrifuged at 4^{0}C at 7,000 rpm for 20 min. The clear supernatant was used as enzyme preparation.

Assay of the Glutamate Synthase

The activity in crude extract was assayed spectrophotometrically following the oxidation of NADH. The assay mixture contained 1.2 ml of 25 mM sodium phosphate buffer (pH 7.5), 0.4 ml of 20 mM 1-glutamine, 0.4 ml of 5 mM 2-oxoglutarate, 1mM EDTA (added in assay buffer), 0.1 ml of 100 mM KCl, 0.6 ml of 1 mM NADH and 0.3 ml of enzyme preparation in a final volume of 3.0 ml. At these concentrations maximum level of the activity was obtained with extract of the tissue studied and the activity was proportional to enzyme concentration in each case. The reaction was started by adding l-glutamine and the decrease in absorbance (which was linear for atleast 5 min) was recorded continuously for 5 min at 340 nM in a Beckman DU-64 spectrophotometer. The amount of NADH was calculated from the standard curve for NADH (Fig 15.1).

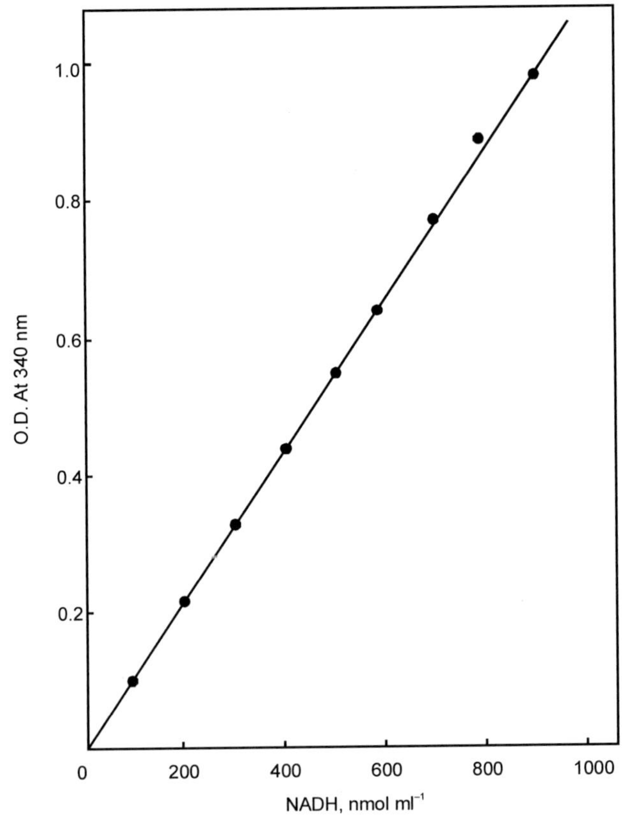

Fig. 15.1 : Calibration curve for reduced nicotinamide adenine dinucleotide (NADH)

Observations

....

....

....

....

Experiment – 16 : Estimate the Amylase Activity in the given Sample by Calculating Units (U) of Enzyme

Principle

Amylases hydrolyse starch and glycogen into sugars. There are two types of amylases: α-amylase and β-amylase. α-amlyase is an endohydrolase which hydrolyses α-1,4-linkage within the chain and breaks these polymers into small fragments. In contrast β-amylase is an exohydrolase and it starts acting from non-reducing end removing two glucose molecules at a time. The hydrolytic action of β-amylase results in increasing reducing power of the starch solution due to the release of a large number of molecules of the reducing sugar (maltose). The reducing sugars can be determined quantitatively by their reaction with DNSA reagent which undergoes reduction to a red-brown coloured product with absorption maxima at 540nm.

Requirements

Reagents

1. Experimental material: use barley seeds which have been kept for germination for 2-3 days at 25^{0}C.
2. DNSA Reagent
3. Sodium Potassium Tartarate (40%): Take 40g of sodium potassium tartarate and dissolve it in small amount of water and make the final volume to 100 ml with D/W.
4. Substrate (1% Starch, w/v): Weigh 1g of starch and add 100ml of boiling water to it. Continue boiling for 2-3 min. Let the solution cool down and then filter it. Use the filtrate as the substrate solution.
5. Sodium phosphate buffer (0.05M, pH 7.0) containing 0.5M NaCl.
6. Standard maltose solution (1 mg/ml): Dissolve 100 mg of maltose in 100ml of D/W.

Equipments: Balance, water bath, spectrophotometer.

Glassware: test tubes, beakers, measuring cylinders, glass pipettes etc.

a) Preparation of Standard curve for maltose

1. Take different aliquot of Standard maltose solution in range of 0.2-mg/ml.

2. Make up final volume to 1ml with D/W.
3. Add 1ml DNSA reagent in each tube and heat all the tubes on boiling water bath for 10 min.
4. Allow to cool mixture under running tap water and add 8ml of D/W to each tube.
5. Measure the absorbance at 540nm against reagent blank.
6. Plot a standard curve of A_{540} vs Concentration of maltose (mg/ml).

Protocol Table

Volume of maltose sol. (ml)	Conc. of maltose sol. (mg/ml)	Volume of water added (ml)	Total volume (ml)	Absorbance (540nm)
0.2		0.8	1.0	
0.4		0.6	1.0	
0.6		0.4	1.0	
0.8		0.2	1.0	
1.0		-	1.0	
-		1.0 (blank)	1.0	

b) Determination of Enzyme activity

Procedure

1. Extraction of the enzyme should be done in cold at 0-4^{0}C. Macerate 500mg of washed germinating seeds in a chilled pestle and mortar with 5ml of 0.05 M phosphate buffer (pH 7.0) containing 0.5 M NaCl.
2. Centrifuge the homogenate at 10,000 X g for 15 min in a refrigerated centrifuge at 4^{0}C. Decant the supernatant and use it as the enzyme preparation.
3. Pipette 1 ml of starch solution and 1 ml of an appropriately diluted enzyme preparation into two tubes (control and experimental).
4. In one of the tubes, which serve as a zero min control, terminate the reaction immediately by adding 2 ml of dinitrosalicylic acid (DNSA) reagent.

5. Incubate the other test tube at 37^0C for 15 min and then stop the reaction by adding 2 ml dinitrosalicylic acid reagent.
6. Keep both the tubes in a boiling water bath for 5 min and then add 1 ml of sodium potassium tartarate to each of the tubes.
7. Cool the tubes under running tap water. Make the volume to 10 min with water.
8. Set the colorimeter to zero absorbance at 540nm with the zero min. control and then record the absorbance of the second tube.

Protocol Table

Test tube	Enzyme(ml)	Substrate(ml)	Buffer(ml)
Enzyme Blank	-	1.0	1.0
Substrate Blank	1.0	-	1.0
Test	1.0	1.0	-

Calculations

$$\text{Unit (U)} = \frac{\text{OD(test)}}{\text{OD(std)}} \times \frac{\text{Conc (std)}}{\text{Aliquot (test)}} \times \frac{1}{\text{Time}}$$

= ___________mg maltose/ml/min

= ___________µ moles of maltose/ml/min

Observations

....

....

....

....

Experiment – 17 : Investigate the Effect of Temperature on Amylase Activity and find out the Temperature Optima for Amylase

Principle

All the enzymes have narrow temperature range for their efficient functioning. To determine enzyme activity in the reaction mixture, assay should be carried out at optimum temperature, where enzyme and substrate binding are more feasible compare to rest of the temperature.

Requirements

Reagents

1. Sodium or potassium phosphate buffer (0.1 mol/l, pH 6.7). 5l
2. Buffered starch substrate (5g/l in phosphate buffer. Mix 5g of soluble starch to a smooth paste with about 50 ml of boiling phosphate buffer solution to boil for 1 min, then cool to room temperature, and dilute to 1 l with buffers solution. 3l
3. Sodium chloride (10 g/l). 1l
4. α-Amylase suitably diluted. 1l
5. Sodium hydroxide (2 mol/l). 1l
6. Sodium potassium tartrate (Dissolve 300 g of this salt in about salt in about 500 ml of water). 500 ml
7. 3,5-Dinitrosalicylic acid (Dissolve 10 g of this reagent in 200 ml of 2 mol/l sodium hydroxide). 200 ml
8. Dinitrosalicylate reagent (Prepare this by mixing solutions 6 and 8 and making up to 1 liter with water).

Glass/Plastic wares: beakers, test tubes, microtips, measuring cylinder

Equipments: Colorimeter, Water baths, micropipette

Procedure

1. Carried out the similar steps 1-4 as mentioned in the experiment 10.
2. Incubate the different reaction mixture tubes at different temperature such as, 20^{0}C, 25^{0}C, 30^{0}C, 37^{0}C and 45^{0}C for 15 min.
3. Stop the reaction by adding 2 ml DNSA reagent.

4. After that the process is remain the same as mentioned in the experiment 10.
5. Plot graph with Enzyme activity vs temperature and determine temperature optima.

Observations

....

....

....

....

Experiment – 18 : Estimation of Catalase Enzyme from Plants

Principle

Catalase activity can be assayed colorimetrically according to method given in Analytical biochemistry (Sinha, 1972). Catalase facilitates the dismutation of H_2O_2 to water and O_2 according to the reaction.

Catalase

H_2O_2-------------------------$H_2O + \frac{1}{2} O_2$

The enzyme plays an important role in association with SOD as well as in photorespiration and glycolate pathway.

Requirements

Reagents : Phosphate buffers, (0.1 M), pH 7.0, potassium dichromate acetic acid (5% $K_2Cr_2O_7$ + glacial acetic acid in 1:3 ratio), H_2O_2 (0.2 M)
Glass/Plastic wares: beakers, test tubes, microtips, measuring cylinder
Equipments: Colorimeter, centrifuge, Water baths, micropipette

Procedure

1. Take 200 mg plant sample and homogenize with 10 ml of phosphate buffer 0.1 M (pH 7.0).
2. Centrifuge at 10,000 rpm at 4^0C for 30 minutes.
3. Collect the supernatant and store at low temperature.
4. Use supernatant for enzyme assay and estimate the enzyme activity as given below:

Reaction Mixture

Test	Blank	Reagents
1.25 ml	-	H_2O_2
0.50 ml	0.50 ml	Enzyme extract
3.25 ml	4.50 ml	Phosphate buffer
5.00 ml	5.00 ml	

5. Take reaction mixture in conical flask and mix rapidly at 37^0C.
6. After 3 minutes withdraw 2.0 ml reaction mixture and add 2.0 ml potassium dichromate acetic acid reagent.
7. Keep on water bath for 10 minutes and record O.D. at 570 nm against blank.
8. Express result as enzyme unit per gram fresh weight or per gram protein basis.

Observations

....

....

....

....

Experiment – 19 : Estimation of Peroxidase Enzyme from Fresh Leaves

Principle

Peroxidase activity can be assayed colorimetrically according to method given in Plant Physiology 34:416 – 418 (Crune & Galston, 1959). The enzyme catalyses the oxidation of a substrate by removal of hydrogen which combines with H_2O_2.

$$AH_2 + H_2O_2 \dashrightarrow 2H_2O + A$$

Requirements

Reagents : Phosphate buffers, (0.1 M), pH 6.0, pyrogallol (0.1 N), H_2O_2 (0.2%)

Glass/Plastic wares: beakers, test tubes, microtips, measuring cylinder

Equipments: Colorimeter, centrifuge, Water baths, micropipette

Procedure

1. Take 200 mg plant sample and homogenize with 10 ml of phosphate buffer 0.1 M (pH 6.0).
2. Centrifuge at 10,000 rpm at 4^0C for 30 minutes.
3. Collect the supernatant and store at low temperature.
4. Use supernatant for enzyme assay and estimate the enzyme activity as given below:

Reaction Mixture

Test	Blank	Reagents
2.0 ml	2.0 ml	Enzyme extract
2.0 ml	3.2 ml	Phosphate buffer
1.0 ml	-	Pyrogallol
0.2 ml	-	H_2O_2
5.2 ml	5.2 ml	

5. Shake the mixture well and keep it at 37^0C on water bath for 10 minutes for the formation of purpurogalin.
6. Measure the activity at 430 nm and express result as enzyme unit per gram fresh weight or per gram protein basis.

Observations

....

....

....

....

Experiment – 20 : Estimation of Superoxide Dismutase Enzyme from Plant Tissue

Principle

Superoxide dismutase occurs in all aerobic organisms and play essential role in the survival of these organisms. It is the most abundant metalloproteins. The enzyme has been localized to the chlorophyll cytosol and thylakoid membrane. They protect the tissues from the deleterious effect of oxygen free radicals formed in various enzyme reactions in which a single electron is transmitted to O_2.

$$O_2 + e^- \longrightarrow O^-_2 \text{ (superoxide)}$$

$$O^-_2 + O^-_2 + 2H^+ \longrightarrow H_2O_2 + O_2$$

Catalase

$$2H_2O_2 \longrightarrow 2H_2O + O_2$$

According to equation vindication mechanism of SOD is intervened with other enzyme and antioxidant reduced the risk of hydroxyl radical foration.

Mechanism of Assay

Superoxide dismutase (SOD) is assayed on the basis of its ability of inhibit the photochemical reduction of nitro blue tetrazolium (NBT) into blue formazon which has a wavelength maximum at 560 nm. SOD essentially inhibited the formation of the blue formazon and could be quantitated on this basis. The chemical reactions involved are:

1. Photoreduction - oxidation reaction between riboflavin and methionine with the flavin being reduced.
2. Generation of superoxide (O_2^-) from reoxidation of reduced flavins in air.
3. a. Reduction of nitro blue tetrazolium (NBT) by superoxide into blue formazan with wavelength maxima at 560 nm.

   b. Inhibition of step 3a by SOD dismutation of the superoxide into hydrogen peroxide and molecular oxygen.

Extraction

1. Weigh about 0.2 g FW leaf sample (cut into small pieces).

2. Add a pinch of acid-washed sand and 2 ml cold phosphate buffer pH 7.0 (0.05M). Grind in clean, dry mortar and pestle in ice bath.
3. Transfer to centrifuge tubes and wash 3x with buffer (1 ml each of the buffer pH 7.0 and pH 7.8 which were used in initial trial extraction a total of 2 ml per washing) to make a total of 8 ml.
4. Centrifuge at 30,000 x g for 30 minutes at 0-4^{0}C (about 15,000 RPM in our refrigerated centrifuge, be sure to put tubes of more or less the same weight in exactly opposite positions to balance). Collect the clear soup and keep at 0-4^{0}C for assay.

Preparation of Solutions

Keep the solutions in reagent bottles properly labeled with the name and concentration (for the buffers, indicate the H as well, use amber bottles for solutions 2 and 3)

1. 0.05 M phosphate buffers (pHs 7.0 and 7.8).
   a. Dissolve 13.4 g $Na_2HPO_4.7H_2O$ (MW = 268) and 6.9 g $NaH_2PO_4.H_2O$ (MW=137.99) in sterile dH_2O to total volume of 1 litre.
   b. Take an aliquot of solution 1a and adjust pH to 7.8. Record the initial pH (about 3.0 ml per 500 m buffer).
   c. Adjust the pH (about of another aliquot of solution 1a as in step 1b but adjust the pH to 7.0). Dissolve 1 g polyvinyl pyrrolidone per 100 ml buffer (1% w/w PVP final concentration).
   d. Keep the buffers in reagent bottles labeled with phosphate pH 7.0 and 7.8, respectively.
2. 0.15 M EDTA. 20 mM methionine, and 0.12 nitro blue tetrazolium (NBT) in buffer pH 7.8.
   a. Weigh 27.92 mg Na_2 EDTA.$2H_2O$ (MW = 32.2), 1.4921 g methionine (MW = 149.21) and 49.06 mg NBT (MW = 8.176) and dissolve together in 100 ml buffer pH 7.8. This stock solution gives the following concentrations: 0.75 mM EDTA, 100 mM methionine, and 0.6 mM NBT.
   b. Aliquot 10.0 ml of solution in step 2a and dilute to a total of 50 ml with buffer pH 7.8. This gives the following

concentration: 0.15 mM EDTA, 20 mM methionine, and 0.12 mM NBT.

c. Use 2.0 ml of solution 2b per 3 ml assay total volume with the following concentrations: 0.1 Mm EDTA, 13 mM methionine, and 0.08 NBT.

3. 12 μM riboflavin in buffer pH 7.8 (to be prepared fresh prior use).

   a. Weigh 4.5 mg riboflavin (MW = 37637) and dissolve in 400 ml buffer pH 7.8. This gives riboflavin concentration of 29.89 μM.

   b. Aliquot 20.0 ml of solution in step 3a and dilute to a total of 50 ml with buffer pH 7.8. This gives riboflavin concentration of 11.96 μM.

   c. Use 0.5 ml of solution 3b (11.96 μM riboflavin) per 3 ml assay total volume with a final concentration of 1.993.

Assay Procedures

1. 3 ml reaction mixture should contain the following

   Note: Nitro blue tetrazolium and riboflavin solutions must be placed in amber bottles to minimize photoreduction.

   a. 0.05 M phosphate buffer (pH 7.8) ml

   b. 13 mM methionine ml

   c. 75 μM nitro blue tetrazolium ml

   d. 100 nM EDTA ml

   e. 0-200 μl enzyme crude extract (component 1a should be added to dilute the enzyme extract)

2. Shake all the tubes and place the sample tubes 30 cm below a light bank consisting of two 30-watt-fluorescent bulbs (reference tube must be placed in the dark).

3. Start the reaction by switching on the light and allow to run for 20 min.

4. Terminate the reaction by switching off the light with the tubes covered with a black cloth. Tubes are then singly wrapped with used carbon paper or aluminum foil.

5. Read absorbance against the blank (reference tube) at 560 nm. Volume of enzyme extract would range from 0 mixture of

which gives ABS_1 (V) to 200 μl (mixtures of which give ABS_2 (V) then onwards).

6. Plot the ration of ABS_1/ABS_2 in the y-axis against volume of the enzyme in the reaction mixture. The volume of enzyme extract producing 50% inhibition of the reaction (that is with ABS_1/ABS_2 ratio 2 of is to be derived from the graph). The volume of extract which gave 50% inhibition to defined to have one unit of SOD activity.

Calculation

Use blank to set instruments ZERO and measure absorbance of reference (V) and test (v) at 560 nm. Calculate enzyme activity as under:

SOD units/ml = [(V/v) – 1] x (dilution factor)

- Dilution factors 1 = 5 (for/ml activity).
- Dilution factor 2 = 40 (for/g fresh weight activity).

Since 0.2 g fresh weight was extracted with 8 ml buffer solution.

Observations

....

....

....

....

Experiment – 21 : Estimation of Ascorbate Peroxidase Enzyme from Plants

Principle

Ascorbate peroxidase for the reduction of hydrogen peroxide in chloroplasts and the primary product of ascorbate peroxidase reaction, monodehydro-ascorbate, and its disproportionation product, dehydroascorbate, are reduced to ascorbate by respective reductases using NADPH and GSH. GSSG is reduced to GSH by glutathione reductases using photoproduced NADPH in thylakoids. The sequence of oxidation reduction reactions involved in peroxide scavenging system is as under:

Ascorbate peroxidase activity can be assayed colorimetrically according to method gives in plant cell physiol (Y. Nakano, K. Asada, 1987).

$$2O^{-2} + aSe + 2H \longrightarrow 2H_2O_2 + DHasc$$

$$2DHasc + 4H_2O \longleftarrow 2H_2O_2 + 2aSc$$

Requirements

Reagents : 1M and 50mM potassium phosphate buffer, (pH 7.8), 10 mM H_2O_2, 1mM and 10 mM ascorbate in sterile dH_2O

Glass/Plastic wares: beakers, test tubes, microtips, measuring cylinder

Equipments: Colorimeter, centrifuge, Water bath, micropipette

Extraction

Take 700 mg plant sample and homogenized with 5 ml of 5 mM potassium phosphate buffer pH 7.8 with 1 mM ascorbate. Centrifuge at 10,000 rpm at 4^0C for 30 minutes. Collect the supernatant and stored at low temperature. Use the supernatant for enzyme assay and estimated activity as given.

Enzyme Assay

Ascorbate peroxidase was measured by decrease in absorbance A) at 290 mM due to ascorbate oxidation. Run a time course at wavelength 290 using 1 ml cuvette with sterile dH_2O, 50 μl 10 mM aqueous ascorbate, 100 μl of crude extract. Then mix thoroughly and cover the sample compartment of the spectrophotometer. Start scan to observe the straight line while continuously scanning add immediately 50 μL of 10 mM H_2O then mix with plastic ladle and cover the sample compartment of the spectrophotometer. Once a straight line is formed after the addition of 10 mM H_2O_2, a cuvette containing the next sample is introduced to replace the older sample curette.

Note :

The total assay volume is each curette must be equal to 1000 μL or 1 ml. in case where the aliquot of crude enzyme extract has to be changed a corresponding change in aliquot of sterile dH_2O must be done in order to maintain the total assay volume of 1000 μl or 1 ml. some time and paper by adjusting the total scanning time to combine 3-4 sample scans or even more if the number of curettes permits, in one piece of graphing paper.

Keep all graphing papers in a folder for data collection. Record the slope (change in OD per unit time). The molar absorptivity constant of oxidized ascorbate product measured by UV method in 2.8 optical density $(mM)^{-1}$ ascorbate $(cm)^{-1}$. Express activity of ascorbate peroxidase in terms of moles ascorbate oxidized per mg protein per minute.

Observations

....

....

....

....

Experiment – 22 : Estimation of Ascorbate Content Determination from Fresh Leaves

Principle

Ascorbate content can be assayed colorimetrically according to method gives in Biol. Chem. J, Roe and Kuether (1943), Agi.Biol.Chem., Shigeoker, Yokata, Nkano and Kitaoka (1979).

Requirements

Reagents

1. 5% metaphosphoric acid: 5 g of metaphosphoric acid in dH_2O to give a total final volume of 100 ml.
2. Ascorbic acid: purpose 1mg per ml standard stock solution. Dissolve 10 mg ascorbate acid in 5% metaphosphoric acid o a final volume of exactly 10 ml.
3. 3mM DCIP: 43.52 mg of 2,6-DCIP sodium salts dissolve in 50 ml of dH_2O.
4. 1% thiourea.
5. 20% H_2SO_4.
6. 10mM DNPH (2,4-di nitro phenyl hydrazine, MW = 198.1). dissolve 99.05 mg of 2,4-DPH in 50 ml of 20% H_2SO_4.
7. 85% H_2SO_4.

Glass/Plastic wares: beakers, test tubes, microtips, measuring cylinder

Equipments: Colorimeter, centrifuge, Water baths, micropipette

Procedure

1. Homogenize leaf sample (1 g per 20 ml extractant, solution 1 above) with 5% metaphosphoric acid in a mortar and pestle.
2. Centrifuge at 15,000 rpm in refrigerated centrifuge for 30 minutes. Then discard the residue and collect supernatant for ascorbate determination.
3. There are two assay to be done at the same time for each sample the total ascorbate assay ant the oxidized ascorbate assay. Both assays are described in steps number 4a and 4b which follow below.

4. a) For total ascorbate , aliquot 1 ml of the supernatant and add 0.5 ml of 3mM DCIP, to oxidized ascorbic acid into dehydroascorbic acid quantitatively. Follow step 4a in the assay of the standard aliquots prepared in solutions 2 above.

   b) For the oxidized ascorbate aliquot 1 ml of the supernatant and 0.5 ml sterile dH_2O.

5. Prepare a reference tube containing 1 ml of the extractant to be used and 0.5 ml of 3mM DCIP to be used in the assay of standards and both the total and oxidized ascorbate samples. Note too that the final volume of the reference tube at this time is 1.5 ml also.

6. Let all the tubes in steps 4 and 5 to stand for 20 minutes after which add 1 ml of 1% thiourea in 5% metasphosphoric acid, solution 4 above to every tube (note that each tube contains a total of 2.5 ml this time).

7. Add 1 ml of 20% H_2SO_4 for the reference tube only. But for the rest tubes (Standard, total ascorbate and oxidized ascorbate) add 1 ml of 10 mM DNPH dissolve in 20% H_2SO_4 to form the 2,4-DNPH derivative of dehydroascorbic acid (each tube must have a total of 3.5 ml this time).

8. All tube will be incubated at 50^0C for 1 hour.

9. After the incubation, they were allowed to cool in ice bath and after the first 15 minutes still on ice bathe add very gradually 2.5 ml of 85% H_2SO_4, solution 7 (addition has to finish in about 1 minute per tube so as not to allow the temperature of the mixture to be raised appreciably as this will charring of sugars or other organic matter which interferes in the assay). This give a total volume off 6.0 ml each tube.

10. When the 2,4-DNPH derivatives of dehydro ascorbic acid is treated with 85% H_2SO_4, a reddish product is formed which absorbs maximally at 500-550 and 350-380 nm. Each tube must be in ice bath for total 30 minutes than proceed to next step.

11. Finally, add 1 ml of 20% H_2SO_4 to tubes for standards, total ascorbate, and oxidized ascorbate. But in the reference tube add 1 ml of 10 mM DNPH dissolved in 20% H_2SO_4 to give a total of 7 ml for each tube.

12. Measure absorbance at 520 nm against the reference. The coloured product is stable for 1 hour.

Observations

....

....

....

....

Additional Reading

Abelsen, J.N. and Simon, M.I. (editors) (1992). Methods in Enzymology. Academic Press Inc., New York.

Alberty et al. (1954). J. Amer. Chem. Soc. 76 :2485-2493.

Boosman, A. and Chilson, O.P. (1976). J. Biol. Chem. 251 : 1847-1852.

English, B.P., Min, W., van Oijen, A.M., Lee, K.T., Luo, G., Sun, H., Cherayil, B.J., Kou, S.C. and Xie, S. (2006). Ever-fluctuating single enzyme molecules: Michaelis-Menten equation revisited. Nature Reviews Chemical Biology, 2: 87-94.

Frey, P. and Hegeman, A. (2007). Enzymatic Reaction Mechanisms. Oxford: Oxford University Press. (Discusses over 100 case studies of enzyme mechanism.)

Kasparek & Pratt (1977). J. Chem. Ed. 54 : 515.

Koshland, D.E. Jr. (1984). Control of enzyme activity and metabolic pathways. Trends Biochem. Sci. 9: 155-159.

Mahn, M.M. and Husic, D.H. (1989). Biochemical Ed. 17 : 94-96.

Nigh (1976). J. chem. Ed. 53 : 668.

Sumner, J.B. and Somers, G.F. (1953). Chemistry and Methods of Enzymes. Academic Press, Inc., New York.

Silverstein et al. (1974). in Spectrometric Identification of Organic Compounds (3rd edition, Wiley), pp. 181-182, 186-187, 189-191.

Teipel et al. (1968). J. Biol. Chem. 243 : 5684.

Weaver & Banaszak (1996). Biochemistry. 35 : 13955-13965.

❑❑❑

Unit -7

Vitamins

Franz Holfmeister's (LT, 1850-1922) definition, "vitamins are substances which are indispensable for the growth and maintenance of the animal organism, which occur both in animal and plants and are present in only small amounts in food", still holds good, although it has been interpreted in various ways.

Fig 7.1 : Source of vitamins

The term vitamin, in its modern sense, usually refers to the substances distinct from major components of food, required in minute quantities (i.e., oligodynamic in nature) and whose absence causes specific deficiency diseases. As the living organisms cannot synthesize most of these compounds, a steady supply of them is an indispensable condition for life. Their ultimate source is the plant or bacterial world.

Fig 7.2 : Vitamin rich our food

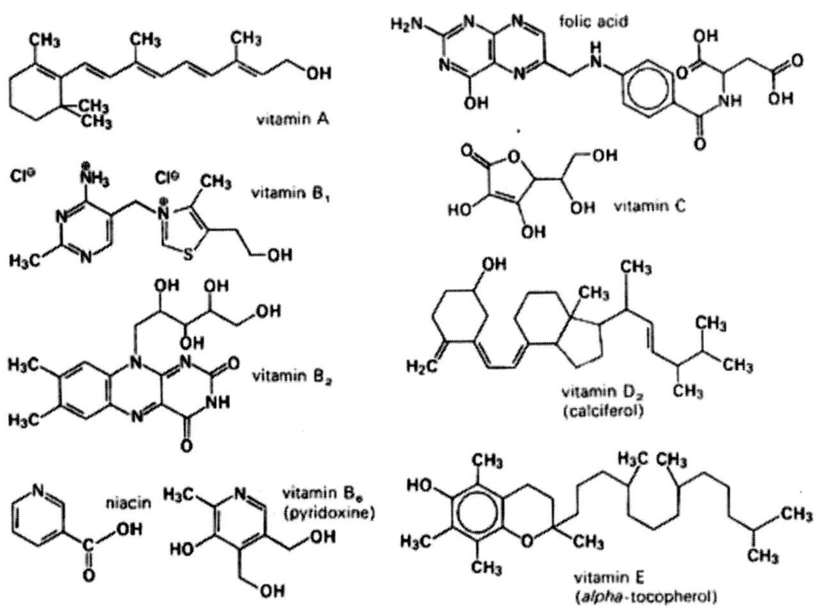

Fig 7.3 : Structure of different vitamins

Fig 7.4 : Vitamins and our food

- *"Knowledge is proud that he knows so much wisdom is humble that he know no more".*

Cowper

Experiment – 1 : Estimation of Ascorbic Acid (Vitamin C) by Volumetric Method

Principle

Ascorbic acid reduces the 2,6-dichlorophenol indophenols dye to a colour less leuco-base. The ascorbic acid gets oxidise to dehydroascorbic acid. Though the dye is a blue coloured compound, the end product is the appearance of pink colour. The dye is pink coloured in acid medium. Oxalic acid is used as the titrating medium.

$$\text{Ascorbic acid} + 1/2\, O_2 \xrightarrow[\text{oxidase}]{\text{ascorbic acid}} \text{dehydroascorbic acid} + H_2O$$

Requirements

Reagents

1. 4% Oxalic acid
2. Dye solution: Weigh 42mg sodium bicarbonate into a small volume of distilled water. Dissolve 52 mg 2,6-dichlorophenol indophenols in it and make up to 200 ml with distilled water.
3. Stock standard solution: Dissolve 100 mg ascorbic acid in 100 ml of 4% oxalic acid solution in a standard flask (1 mg/ml).
4. Working Standard: Dilute 10 ml of the stock solution to 100 ml with 4% oxalic acid. The concentration of working standard is 100 µg/ml.

Glass/Plastic wares: beakers, test tubes, microtips, measuring cylinder, conical flask.

Equipments: Centrifuge, Water bath, micropipette.

Procedure

1. Pipette out 5 ml of the working standard solution into a 100 ml conical flask.
2. Add 10 ml of 4% oxalic acid and titrate against the dye (V_1 ml). End point is the appearance of pink colour which persists for a few minutes. The amount of the dye consumed is equivalent to the amount of ascorbic acid.

3. Extract the sample (0.5 – 5 g depending on the sample) in 4% oxalic acid and make up to a known volume (100 ml) and centrifuge.
4. Pipette out 5 ml of this supernatant, add 10 ml of 4% oxalic acid and titrate against the dye (V_2 ml).

Calculation

Amount of ascorbic acid (mg/100g sample)

$$= \frac{0.5 \text{ mg}}{V_1 \text{ ml}} \times \frac{V_2}{15 \text{ ml}} \times \frac{100 \text{ ml}}{\text{Wt. of the sample}} \times 100$$

Note :

Acetic-metaphosphoric acid mixture (15g of meta phosphoric acid dissolved in a mixture of 50 ml of glacial acetic acid and 450 ml of water) may also be used instead of 4% oxalic acid.

Observations

....

....

....

....

Experiment – 2 : Estimation of Ascorbic acid (Vitamin C) by Colorimetric Method

Principle

Ascorbic acid or vitamin C is a sugar acid and an antiascorbutic. It is required for normal formation of connective tissue collagen, specifically for hydroxylation of certain proline and lysine residues. It is also required for normal iron metabolism. It is a strong reducing agent losing hydrogen atom readily to become dehydroascorbic acid which has also vitamin C activity. Ascorbic acid acts as an antioxidant and protects the cell membrane from the toxic action of powerful oxidizing agents. Ascorbic acid is found abundantly in berries, fresh fruits like citrus, guavas, chilies and green leafy vegetables. Ascorbic acid is first dehydrogenated by bromination. The dehydroascorbic acid is then reacted with 2,4 dinitrophenyl hydrazine to form osazone and dissolved in sulphuric acid to give an orange – red color solution which is measured at 540 nm.

Requirements

Reagents

1. 4% Oxalic acid.
2. 0.5 N Sulphuric acid
3. 2% 2,4 Dinitrophenyl hydrazine (DNPH) reagent : Dissolve by heating 2 g DNPH in 100 ml 0.5N H_2SO_4. Filter and use.
4. 10% Thiourea solution.
5. 80% Sulphuric acid.
6. Bromine water : Dissolve 1-2 drops of liquor bromine in approximately 100 ml cool water.
7. Ascorbic acid stock solution: Dissolve 100 mg ascorbic acid in 100 ml of 4% oxalic acid solution in a standard flask (1 mg/ ml).

Glass/Plastic wares: beakers, test tubes, microtips, measuring cylinder.

Equipments: Spectrophotometer, Water baths, micropipette.

Procedure

Extraction

1. Grind 0.5 – 5 g of sample material either mechanically or using a pestle and mortar in 25-50 ml 4% oxalic acid solution.

2. Centrifuge or filter and collect the liquid.
3. Transfer and aliquot (10 ml) to a conical flask and add bromine water drop wise with constant mixing (the enolic hydrogen atoms in ascorbic acid are removed by bromine).
4. When the extract turns orange yellow due to excess bromine, expect it by bowling in air.
5. Make up to a known volume (25 or 50 ml) with 4% oxalic acid solution.
6. Similarly, convert 10 ml stock ascorbic acid solution into dehydro form by bromination.

Estimation

7. Pipette out 10-100 μg standard dehydroascorbic solution into a series of tubes.
8. Similarly, pipette out different aliquots (0.1 ml - 2 ml) of brominated sample extract.
9. Make up the volume in each tube to 3 ml by adding distilled water.
10. Add 1 ml of DNPH reagent followed by 1-2 drops of thiourea to each tube.
11. Set a blank as above but with water in place of ascorbic acid solution.
12. Mix the contents of the tubes thoroughly and incubate at 37^0C for 3 h.
13. After incubation dissolve the orange-red osazone crystals formed by adding 7 ml of 80% sulphuric acid.
14. Measure absorbance at 540 nm.
15. Plot a graph of ascorbic acid concentration versus absorbance and calculate the ascorbic acid content in the sample.

Note :

Liquor bromine can cause burns. Pre chill the ampoule containing bromine prior to cut open it.

Observations

....

....

....

....

Experiment – 3 : Estimation of Vitamin A

Principle

Vitamin A is also a fat-soluble vitamin which occurs in two common forms, vitamin A_1 or retinol the form most common in mammalian tissues and marine tissues and vitamin A_2 or retinal common in fresh water fishes. Vitamin A_1 or retinol is the vitamin involved in vision. Vitamin A aldehyde, retinal is present in rod cells of the retina combined with a protein (opsin) as rhodopsin. Deficiency of vitamin A leads to impairment of vision, the first stage is night blindness, the inability to see in dim light. Retinol is found only in animal sources like cod liver and other fish liver oils. In plants, it occurs in the form of precursors or provitamins, the carotenoids like β-carotene present in carrots, green leafy vegetables, etc. When antimony trichloride reacts with vitamin A in chloroform solution a blue colour develops. If the solution containing the vitamin for analysis is freed from saponifiable material and an appropriate color filter is used on photoelectric colorimeter this reaction appears fairly reliable for vitamin A. The vitamin A is estimated from the colorimeter readings by use of a standard curve.

Requirements

Reagents

1. Chloroform : Wash 3 times with an equal volume of water, dry over anhydrous sodium sulfate, distils, and stroe over anhydrous sodium sulfate.
2. Antimony trichloride solution : Weigh an unopened bottle of antimony chloride (25-30g), it, and transfer to a glass-stoppered wide-mouthed, amber-coloured bottle containing 100ml of chloroform.
3. Reweigh the opened bottle and obtain weigh of antimony trichloride added to the chloroform by difference. This solution should be filtered if turbid.
4. Alcoholic potassium hydroxide solution (0.5M) : Dissolved about 35g of potassium hydroxide in 20ml of water and add sufficient alcohol to make 1000 ml of solution.
5. Aqueous potassium hydroxide solution (0.5M).
6. Diethyl ether : Freshly distil over sodium hydroxide pellets.

7. Vitamin A standard : A chloroform solution containing 100 U.S.P. Units of vitamin A/ml. This is prepared by dissolving a weighed amount of distilled vitamin A esters in chloroform.

Glass/Plastic wares: beakers, test tubes, microtips, measuring cylinder.

Equipments: Colorimeter, Water bath, micropipette.

Procedure

1. Weigh an amount of fat or oil containing at least 50 U.S.P. Units into a saponification flask, add the alcoholic potassium hydroxide (10 I/g of sample) and attach to a reflux condenser.
2. Heat on a water bath for 30 min.
3. Wash the condenser with water, cool, dilute with 50-100ml of water, and transfer to a separatory funnel.
4. Extract 4 times with 50-100ml of ether.
5. Combine the ether extracts, pour two 50ml portions of water through the combined ether extracts and discard the water without shaking.
6. Wash the ether extract with 50ml of the 0.5M aqueous potassium hydroxide, shaking gently.
7. Allow to separate, draw off aqueous layer, and discard.
8. Wash with 50 ml portions of water until free of alkali.
9. Allow the ether extract to stand for 5min, discard separated water and filter through 305g of anhydrous sodium sulfate placed on filter paper in a funnel into a 250ml flask.
10. Rinse the separatory funnel with small portions of ether and add the rinsings to the 250ml flask.
11. Evaporate the ether to dryness on a water bath, removing the flask from direct heat towards the end of the evaporation (viscous oily residue).
12. Take up the residue immediately in chloroform adjusting the concentration to 5-15 U.S.P. Units/ml.
13. Transfer 2ml of chloroform to a colorimeter tube or cuvette and add 9ml of the antimony trichloride reagent with the aid of fast delivery pipette and zero the instrument using the blank.

14. To another tube or cuvette containing 1ml of the unknown solution and 1ml of chloroform add 9ml of the antimony trichloride solution.
15. Stopper the tube immediately, swirl, and read the absorbance on the instrument set at 620nm (A) (the reading should be made within 3-6 see after the addition of antimony trichloride solution).
16. To another tube or cuvette add 1ml of the unknown solution, 1 ml of a known vitamin A solution in chloroform approximately equal in concentration to that of the unknown, and treat as above (B).

Calculation

$$\text{U.S.P. Units/ml} = \frac{A}{B-A} \times \text{Concentration of standard(U.S.P. Units/ ml)}$$

$$\text{Vitamin A/g sample} = \text{U.S.P. Units/ml unknown x } \frac{\text{Final volume}}{\text{Sample weight}}$$

Note :

If the unknown is coloured, a blank correction is made by measuring the absorbance at 620nm of 1ml of unknown plus 10ml of chloroform.

Observations

....

....

....

....

Experiment – 4 : Estimation of Vitamin D_2 and D_3

Principle

Vitamin D, calciferol is fat-soluble and has two important forms- D_2 (ergocalciferol) found in plants and D_3 (cholecalciferol) found in animals. Cholecalciferol is found in fish liver oil and milk but can be formed from the precursor sterol (7-dehydrocholesterol) in skin under the influence of ultra-violet light. The active form of vitamin D_3 is 1, 25-dihydroxycholecalciferol (calcitriol) which is required for the normal formation of bone. The vitamin also increases the calcium absorption into the intestinal cell and brings about normal calcification of bone matrix. Deficiency of the vitamin results in rickets in children. By proper control of the composition of an antimony trichloride reagent it can be shown that saturated sterols do not react therewith as chromogens but unsaturated steroids do. Sterols with one double bond in ring B give a yellow color with a low absorption maximum at 500nm sloping from the violet towards the red. Sterols with two double bonds in ring B show a maximum at 510 to 515nm while vitamins D_2 and D_3 have the very sharp and high maximum absorption at 500nm.

Requirements

Reagents : It is important that the reagents be free from alcohol, water and pentavalent antimony.

1. Chloroform: Wash reagents chloroform with equal volume of water seven times. Then shake the chloroform with an excess of P_2O_5 and rapidly pass through a filter. Distil the chloroform, discarding the first turbid distillate and the last 10%.
2. Antimony trichloride reagent: In 100 ml of the purified chloroform dissolve 15-22g reagent grade antimony trichloride by warming the mixture to 35-45^{0}C. Filter and add 2ml of Merck's redistilled acetyl chloride per 100 ml solution. To this add a small amount of 30 mesh zinc, tin or antimony in order to reduce any pentavalent antimony.
3. Standard solution of vitamin D_2 or D_3 : Prepare a solution in the reagent grade chloroform containing 200µg crystalline D_2 or D_3 per 100ml. dilute this solution 10 fold with purified chloroform as needed and use aliquots thereof for standards. Add extra chloroform to make the volume 1ml.

Glass/Plastic wares: beakers, test tubes, microtips, measuring cylinder

Equipments: Colorimeter, Water baths, micropipette

Procedure

1. Follow the steps from 1 to 12 as described under the method of 'Vitamin A' estimation.
2. Transfer 1ml of chloroform solution containing vitamin D to a colorimeter tube and add 24ml of the antimony trichloride reagent (the colour develops to its maximum at room temperature in 30 sec and is stable for 4 or 5 min).
3. Measure the color intensity at 500nm.
4. Calculate the amount of vitamin D_2 and D_3 are identical.

Observations

....

....

....

....

Experiment – 5 : Estimation of Tocopherol (Vitamin E)

Principle

α-Tocopherol is fat-soluble and the most active and abundant type of vitamin E. The vitamin prevents oxidation of unsaturated fatty acids, that is, it acts as an antioxidant. Deficiency of vitamin E results in destruction of erythrocytes, muscular dystrophy (wasting of muscle) and kidney degeneration. It occurs in plant oils especially in those cotton seed, wheat germ, rice bran and soybean.

If extracts of tissues are properly prepared by appropriate solvents and absorbents and the α α′-dipyridine, ferric chloride, glacial acetic acid reagents are employed under well standardized conditions reliable colorimetric estimations of tocopherols in tissue extracts can be made. It is important to avoid exposure of the extracts to light. Operate in a dark room as much as possible and use amber glassware. If amber glassware is not available one can cover the regular glassware with orange (Tango) cellophane.

Requirements

Reagents

1. Special solvent: Prepare a mixture of 8.5 volumes Skelly Solve B (SSB) with 10 volumes redistilled absolute ethanol. The Skelly Solve B should first be purified by shaking in a separatory funnel with conc. H_2SO_4, then washing the petroleum ether layer well with 10% NaOH solution and water. Finally distil.
2. Purified super-sorb (florisil): The material can be obtained from the Floridin Company, Warren, Pa.
3. Reagent benzene.
4. αα′-Bipyridine reagent: Dissolve 250mg ferric chloride and 500 mg αα′-bipyridine in 1 liter glacial acetic acid.
5. Skelly Solve E, 117-127^{0}C (SSE)
6. 85% H_2SO_4
7. 2% KOH in water.

Glass/Plastic wares: beakers, test tubes, microtips, measuring cylinder

Equipments: Colorimeter, Water baths, micropipette

Procedure

1. Take 10-25g of material (muscle) and hash in the warming blender with approximately 200ml of special for 5 min.
2. Centrifuge, separate and treat the residue in the same way three times.
3. To the combined add sufficient water to make about a 40% concentration of alcohol (that is, for every 100ml of absolute alcohol represented by the mixture recovered add 60ml distilled water).
4. Shake gently in a separatory funnel. If an emulsion forms again add alcohol to break the emulsion.
5. Separate and extract the alcohol-water phase four times with Skelly Solve B (SSB).
6. Combine the petroleum ether extracts and evaporate to dryness under diminished pressure in an atmosphere of nitrogen.
7. Dissolve the residue in 15ml of reagent grade benzene. Pass 5ml aliquots through separate 30 x 12 mm adsorption columns of the purified super-sorb and wash each column with four 5ml portions of benzene (it is important to remove cholesterol hence the treatment with super sorb).

Treat the benzene filtrates as follows

8. Add 10ml of Skelly Solve E, 117-127^{0}C (SSE), transfer to a glass stoppered centrifuge tube, add 2ml of the 85% H_2SO_4 shake and centrifuge for 2min.
9. Decant the supernatant as closely as possible to a similar tube, add 5ml of the 2% KOH, shake and centrifuge for 20 min.
10. To 5ml of the supernatant add 20ml of the α α'-bipyridine reagent. Mix well and after 30 min compare in the colorimeter with standards of 10-100µg tocopherolin 5ml of Skelly Solve E and correct for appropriate blanks on reagent, etc., in the range of 485-50nm.

Note :

1. *The procedure above is not satisfactory for liver extracts because the vitamin A interferes in the color development. However, if the shaking with sulphuric acid, NaOH etc., is done before the floridin treatment and the latter done as usual and repeated after evaporation under nitrogen, the final benzene filtrate after evaporation under nitrogen and redissolving in SSE develops the proper color with the reagent.*

Observations

....

....

....

....

Experiment – 6 : Estimation of Pyridoxine (B_6)

Principle

Pyridoxine and two other closely related compounds pyridoxal and pyridoxamine constitute vitamin B_6. All the three forms are found in natural sources and are interconvertible. The active coenzyme forms of vitamin B_6 are pyridoxal phosphate (PLP) and pyridoxamine phosphate which are involved in transamination and decarboxylation reactions in amino acid metabolism. The deficiency of pyridoxine results in weight loss, anemia and convulsions in children. Good sources of pyridoxine are wheat grains, egg yolk and meat. The tissue is extracted and hydrolysed by acid and enzyme action. The proteins are removed by tungstic acid precipitation and the pyridoxine adsorbed on Super filtrol at pH 3. The pyridoxine is selectively eluted by alkaline alcohol and converted into an azo dye which is estimated colorimetrically.

Requirements

Reagents

1. Chloroform: Wash 3 times with an equal volume of water, dry over anhydrous sodium sulfate, distil, and store over anhydrous sodium sulfate.
2. Antimony trichloride solution: Weigh and unopened bottle of antimony chloride (25-30g), open it, and transfer to a glass-stoppered wide-mouthed, amber-coloured bottle containing 100ml of chloroform.
3. Reweigh the opened bottle and obtain weight of antimony trichloride added to the chloroform by difference. This solution should be filtered if turbid.
4. Alcoholic potassium hydroxide solution (0.5M): Dissolve about 35g of potassium hydroxide in 20ml of water and add sufficient alcohol to make 1000ml of solution.
5. Aqueous potassium hydroxide solution (0.5M).
6. Diethyl ether: Freshly distil over sodium hydroxide pellets.
7. Vitamin A standard: A chloroform solution containing 100 U.S.P. Units of vitamin A/ml. This is prepared by dissolving a weighed amount of distilled vitamin A ester in chloroform.

Glass/Plastic wares: beakers, test tubes, microtips, measuring cylinder

Equipments: Colorimeter, Water bath, micropipette

Procedure

1. Weigh an amount of fat or oil containing at least 50 U.S.P. Units into a saponification flask, add the alcoholic potassium hydroxide (10ml/g of sample) and attach to a reflux condenser.
2. Heat on a water bath for 30 min.
3. Wash the condenser with water, cool, dilute with 50-100ml of water, and transfer to a separatory funnel.
4. Extract 4 times with 50-100ml of ether.
5. Combine the ether extracts, pour two 50 ml portions of water through the combined ether extracts and discard the water without shaking.
6. Wash the ether extract with 50ml of the 0.5M aqueous potassium hydroxide, shaking gently.
7. Allow to separate, draw off aqueous layer, and discard.
8. Wash with 50ml portions of water until free of alkali.
9. Allow the ether extract to stand for 5 min, discard separated water and filter through 305g of anhydrous sodium sulfate placed on filter paper in a funnel into a 250ml flask.
10. Rinse the separatory funnel with small portions of ether and add the rinsings to the 250ml flask.
11. Evaporate the ether to dryness on a water bath, removing the flask from direct heat towards the end of the evaporation (viscous oily residue).
12. Take up the residue immediately in chloroform adjusting the concentration to 5-15 U.S.P. Units/ml.
13. Transfer 2ml of chloroform to a colorimeter tube or cuvette and add 9ml of the antimony trichloride reagent with the aid of a fast delivery pipette and zero the instrument using the blank.
14. To another tube or cuvette containing 1ml of the unknown solution and 1ml of chloroform add 9ml of the antimony trichloride solution.
15. Stopper the tube immediately, swirl, and the read the absorbance on the instrument set at 620nm (A) (the reading should be made within 3-6 see after the addition of antimony trichloride solution).

16. To another tube or cuvette add 1 ml of the unknown solution, 1ml of a known vitamin A solution in chloroform approximately equal in concentration to that of the unknown, and treat as above (B).

Calculation

$$\text{U.S.P. Units/ml} = \frac{A}{B-A} \times \text{Concentration of standard(U.S.P. Units/ ml)}$$

$$\text{Vitamin A/g sample} = \text{U.S.P. Units/ml unknown} \times \frac{\text{Final volume}}{\text{Sample weight}}$$

Note :

If the unknown is coloured, a blank correction is made by measuring the absorbance at 620nm of 1ml unknown plus 10 ml of chloroform.

Observations

....

....

....

....

Experiment – 7 : Estimation of Thiamin

Principle

Thiamin is also known as vitamin B_1. Its deficiency causes beri-beri. One form of the disease known as "dry" beri-beri is characterized by loss of appetite, wasting of muscle and neuritis (tingling and numbness in legs and hands). The second form "wet" beri-beri leads to edema and cardiac impairment including heart failure. The biologically active form of thiamin is thiamin pyrophosphate (TPP) which serves as the coenzyme for oxidative decarboxylation of α-keto acids in carbohydrate metabolism. It occurs in outer layers of grains of many plants including cereals. The practice of using polished rice or refined wheat flour removes much of the vitamin. Thiamin is water-soluble and overcooking may leach out or destroy this vitamin in the food sources. Among a number of methods available for the estimation of thiamin, flurormetric method is easier and described here. Alkaline potassium ferricyanide oxidizes thiamin to thiochrome which is a fluorescent compound. The thiochrome is extracted in isobutyl alcohol and measured in a fluorimeter.

Requirements

Reagents

1. 15% NaOH
2. % Potassium ferricyanide
3. Isobutyl alcohol
4. Sodium sulphate (anhydrous)
5. Approximately 0.1N H_2SO_4
6. Standard thiamin hydrochloride

Glass/Plastic wares: beakers, test tubes, microtips, measuring cylinder

Equipments: Fluorimeter, Water bath, micropipette

Stock Solution

Dissolve 50mg thiamin hydrochloride in 500ml of about 0.1N sulphuric acid containing 25% alcohol (100μg/ml). Store this solution in a brown bottle in a refrigerator.

Working Standard Solution

Prior to experiment, dilute 5ml of the stock solution to 100ml with 0.1N sulphuric acid and again dilute 5ml of the second solution to 100ml with 0.1N H_2SO_4 (0.25µg/ml) and use.

Procedure

Extraction of Thiamin

1. Weigh accurately 5g finely ground sample into a 250ml conical flask in duplicate.
2. Slowly add 100ml of 0.1N H_2SO_4 without shaking, stopper the flask and let it stand overnight.
3. Next morning shake vigorously and filter through Whatman No. 1 filter paper, discarding the first 10 to 15 ml of filtrate.

Oxidation

4. Pipette out 10ml of the extract in duplicate into 100ml separating funnels.
5. Pipette out 10ml of working standard (in 4-5 replicates).
6. Add 3ml of 15% NaOH into each separating funnel immediately followed by four drops (0.2ml) of ferricyanide solution.
7. Shake gently for exactly 30 seconds.
8. Add 15ml of isobutanol rapidly from a quick delivery burette or a measuring cylinder.
9. Stopper immediately, shake vigorously for 60 sec and allow the layers to separate.
10. Drain of the bottom layer carefully and discard it.
11. Add one spatula full of sodium sulphate directly into the separating funnel, stopper and swirl gently to clarify the extract. If the extract is not clear, add a little more Na_2SO_4 and clarify.
12. Collect the clear extract from the top using a Pasteur pipette into a clean dry test tube.
13. Prepare a set of sample blanks by pipetting out 10ml of the extract and follow the above procedure but omit addition of ferricyanide.

14. Prepare a blank for the standard (in duplicate) separately.
15. Select suitable primary (366nm) and secondary filters as per the make of the fluorimeter.
16. Set fluorimeter by initially adjusting the standard to 100. Then read the sample blank and sample. Since the light intensity sometimes changes progressively it is advised to take 5-6 readings.

Calculation

μg Thiamin content in 100g sample

$$= \frac{0.25\times10}{a-b}\times\frac{(c-d)\times100}{10}\times\frac{10}{5}$$

Where

a = reading of standard = 100

b = reading of standard blank = 0

c = reading of standard sample and

d = reading of standard sample blank

Note :

1. *Since the standard has to be read a number of times during the measurement it is convenient to oxidize the standard in 4 or 5 replicates and to combine all the oxidize the standard in one conical flask. Portions of this may be read at a time and discarded.*
2. *Thiochrome is rapidly destroyed by UV light, so the solution once exposed should not be read again. It should be discarded.*
3. *Follow the instructions given in the manual for operating the fluorimeter.*

Observations

....

....

....

....

Experiment – 8 : Estimation of Riboflavin

Principle

Riboflavin also known as vitamin B_2 in the form of its prosthetic groups flavin mononucleotide (FMN) and flavin adenine dinucleotide (FAD) serves as a carrier of hydrogen and electrons. Riboflavin is less soluble in water than thiamin but more stable to heat in acid or neutral medium. In solution, it is destroyed when exposed to bright light. Deficiency of riboflavin leads to painful fissures at the corners of the mouth (cheilosis), eye damage and dermatitis of the face (shark skin). Important sources of riboflavin are milk, legumes, cereals and green leafy vegetables. Estimation of riboflavin by fluorometric method is described here.

Riboflavin fluoresces at wavelength 440 to 500nm. The intensity of fluorescence is proportional the concentration of riboflavin in dilute solution. The riboflavin is measured in terms of the difference in fluorescence before and after chemical reduction.

Requirements

Reagents

1. 0.1N H_2SO_4
2. 2.5M sodium acetate
3. 4% Potassium permanganate (prepare fresh every time)
4. 3% Hydrogen peroxide
5. Riboflavin standard:
   (a) Dry standard riboflavin over P_2O_5 in a desiccators for 24h.
   (b) Dissolve 50mg in 1500ml water and 2.4ml glacial acetic acid in a two litre flask. Warm, cool and make up to volume.
   (c) Store under toluene in amber bottles in a refrigerator (25μg/ml).
   (d) Dilute 20ml of the above riboflavin standard solution to 50ml with water. For working solution, dilute 10ml of the second riboflavin standard solution with water to 100ml. prepare this dilution fresh. Protect from light (1ml = μg riboflavin).
6. Sodium hydrosulphite (solid).

Glass/Plastic wares: beakers, test tubes, microtips, measuring cylinder

Equipments: Fluorimeter, Water bath, micropipette

Procedure

1. Weigh accurately 2-5 g of thoroughly mixed and ground sample into a 250ml conical flask.
2. Add 75ml 0.1N H_2SO_4 and mix.
3. Autoclave at 15lbs for 30 min or immerse the flask in boiling water for 30 min. shake the flask for every 5 min. Let is cool to room temperature.
4. Add 5ml 2.5M sodium acetate solution, mix and let stand for at least 1h. The solution is now at approximately pH 4.5.
5. Transfer to a 100ml volumetric flask and make the volume with water. While transferring, wash the conical flask completely to transfer the material without any loss.
6. Filter through medium fast paper such as Whatman No. 2 or No. 4 discarding the first 10-15ml.
7. In test tubes of one inch diameter marked A and B with stirring bars carry out oxidation as follows:

	Low/Blank sample		**High/Blank sample**	
	Tube A	Tube B	Tube A	Tube B
Sample solution (ml)	10	10	10	10
Standard solution (ml)	1	-	1	-
Water (ml)	1	2	-	1.0
*Potassium permanganate (4%) (ml)	0.5	0.5	1.0	1.0
Time lapse (min)	2	2	4	4
**Hydrogen peroxide (3%) (ml)	0.5	0.5	1.0	1.0

*Stir samples after addition of permanganate

**Shake after adding peroxide until foaming is negligible

8. Set the fluorimeter so that glass standard or sodium fluorescein solution gives suitable galvanometer deflection, as direction in the instruction manual of the instrument. Determine fluorescence of solutions A and B. Do not expose the solution for more than 10 seconds.

9. To solution B in the curvette, and 20mg sodium hydrosulphuric, stir and note the blank fluorescence (c). Do not use reading (c) taken after colloidal sulphur begins to form.

Calculation

$$\text{Riboflavin } (\mu g/100g) = \frac{B-C}{A-B} \times \frac{R}{S} \times \frac{V}{V_i} \times 100$$

Where,

A = reading of the sample plus riboflavin standard

B = reading of sample plus water

C = reading after adding sodium hydrosulphite

R = standard riboflavin added = $\mu g/V_1$ of sample solution

V = original volume of sample solution = ml

V_1 = volume of sample solution taken for measurement (10ml)

S = sample weight (g)

In this dilution R = 1; V = 100 and $V_1 = 10$

Observations

....

....

....

....

Additional Reading

Briggs, M.H. (editor) (1984). Recent vitamin research. CRC Press Boca Ruton, Fla.

Dyke, S.F. (1965). The chemistry of the vitamins. Interscience Publishers, Inc., New York.

Meiklcjohn, A.P.C. (1953). Physiology and biochemistry of ascorbic acid. Vitamins and Hormones. 11: 62-96.

Wagner, A.F. and Falkers, K. (1964). Vitamins and coenzymes. Interscience Publishers, Inc., New York.

❑❑❑

Unit -8

Pigments

The pigments are energy-rich organic compounds. The potential chemical energy of these compounds comes from the light energy. The light enrgy to be effective in photosynthesis must be absorbed by a suitable pigment. This vital role is performed by a suitable pigment. This vital role is performed by the green pigment. Chlorophyll, carotenoid (carotenes and xanthophylls), anthocynin and phycibilins.

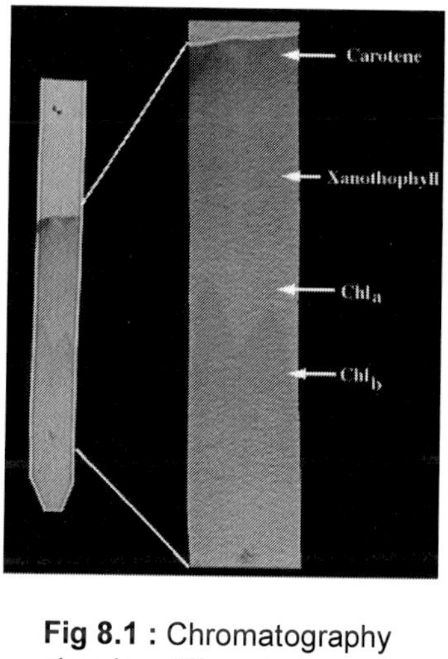

Fig 8.1 : Chromatography showing different pigments

Fig 8.2 : Pigments in plants

There are at least seven types of chlorophylls known chlorophylls a, b, c, d and e, bacteriochlorophyll and bacterioviridin.

All these chlorophyll molecules contain tetrapyrrole skeleton formed into a ring with an atom of magnesium in the centre of the ring 'A' so called pyrrole molecule contains a skeleton. The chemical skeleton of chlorophyll a of five atoms, four carbon and one nitrogen and the five are arranged in a ring.

Carotenoids are widely distributed in plants. They are located in chloroplasts and chromoplasts. They are composed of two six-membered rings with a hydrocarbon chain stretched between them.

The carotenoids includes orange carotenes and yellow xanthophylls. The colour of leaves is modified in certain plants due to the presence of purple pigment called anthocynin. They are present in flowers fruits and fall foliage etc. In blue green algae some additional pigments known as phycobilins are present. They are also tetrapyrroles like chlorophylls but the four joined pyrrole rings form a straight chain.

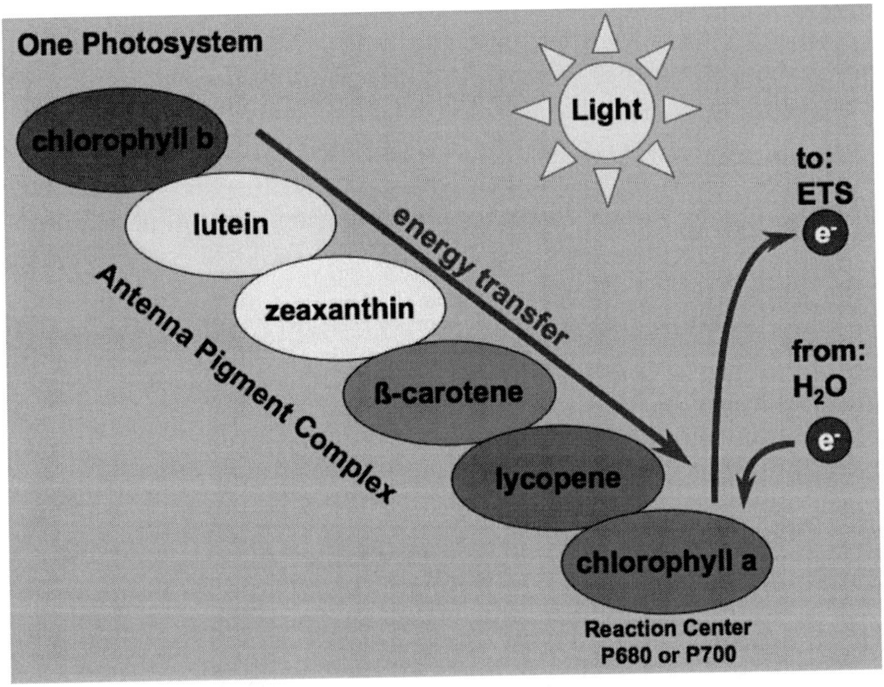

Fig 8.3 : Energy transfer from pigments to chlorophyll for photosynthesis

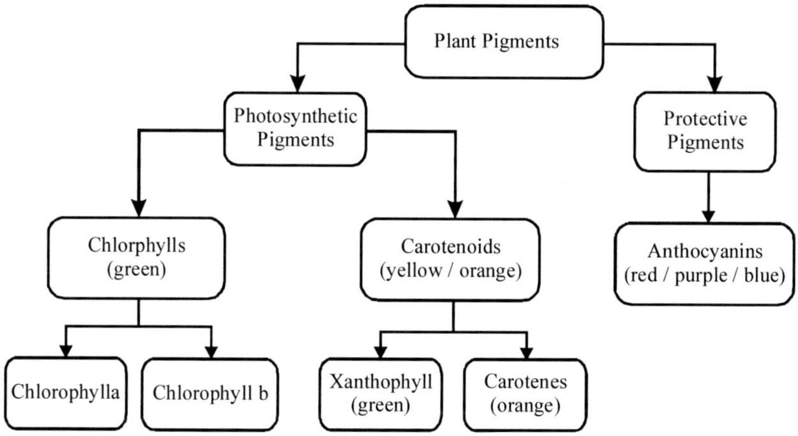

Fig 8.4 : Plant pigments

- *"The scientist is not a person who gives the right answer, he's one who asks the right questions."*

  Claude Levi- Strauses

- *"Winning isn't everything but wanting to win is."*

  Vince Lomardl

Experiment – 1 : Estimation of Total Carotenoids from Plant Tissue

Principle

Carotenoids are tetra-terpenoid (C_{40}) compounds widely distributed in plants. They function as necessary pigments in photosynthesis and as colouring matters in flowers and fruits. Their levels are altered during physiological and pathological conditions especially by viruses. Some of the commonly occurring carotenoids are simple unsaturated hydrocarbons based on lycopene and their oxygenated derivatives known as xanthophylls. β-Carotene is the most common pigment in this group found in higher plants.

The total carotenoids are extracted and partitioned in organic solvents (acetone or methanol) on the basis of their solubility. Carotenoids that are bound as esters are hydrolyzed using aqueous 60% KOH. The amount of carotenoids present in sample is estimated colorimetrically at 450nm using β-carotene as a standard. A method to extract total carotenoids from fresh plant materials and their determination is described below.

Requirements

Reagents : Acetone or methanol (distilled), Peroxide-free ether, Ethanol, Aqueous 60% KOH, Standard β-carotene solution

Glass/Plastic wares : Buchner funnel, Beakers, measuring cylinder, test tubes.

Equipments: Hot water bath, colorimeter, pestle mortar, weighing balance, filter paper (Whatman No. 42.)

Procedure

Sample Extraction

1. Cut the fresh plant material and grind a known amount (2g) in a mortar with 20ml of either distilled acetone or methanol.
2. Filter on a Buchner funnel through Whatman No. 42 filter paper.
3. Repeat the extraction until the tissue is free from pigments.

4. Pool the filtrates and partition thrice with equal volume of peroxide-free ether using separatory funnel (add water if necessary to produce two layers during initial ether extraction).
5. Evaporate the combined ether layers (which contain carotenoids) under reduced pressure at 35^0C in a rotary evaporator or in a hot water bath.
6. Dissolve the residue in minimum quantity of ethanol.
7. Add 60% aqueous KOH at the rate of 1ml for every 10ml, of the ethanol extract to saponify (this will remove chlorophylls, lipids and also cleave the esterified carotenoids).
8. Keep the mixture in the dark in the presence of nitrogen or boil it for 5-10 min or leave it overnight at room temperature.
9. Add equal volume of water and partition twice with ether.
10. Evaporate the combined ether layers as before and dissolve the residue in minimum volume of ethanol.
11. Measure the absorbance of this solution at 450nm and calculate the carotenoid content (mg/100g) in the sample using a calibration curve prepared against a high purity β-carotene.

Notes :

1. *Alcohol may be used instead of acetone or methanol for extraction.*
2. *Use a blender for homogenization of large samples.*
3. *An estimate of total carotenoid content of crude plant extracts containing negligible amounts of chlorophylls may be obtained directly as follows:*

Make suitable dilution of the acetone or methanol extract of the plant tissues and measure the absorbance at 450 and 670nm in a colorimeter. If A_{450} is 10 times that of A_{670}, the carotenoids present in the extract may be estimated from the A_{450} values using the following formula:

$$C = \frac{D \times V \times f \times 10}{2500}$$

Where, C = Total amount of carotenoids (mg)

D = Absorbance at 450 nm in a 1.0 cm cell

V = Volume of the original extract in ml

f = Dilution factor and

2500 = Average extinction coefficient of the pigments

Observations

....

....

....

....

Experiment – 2 : Estimation of Chlorophyll from Plant Tissue

Principle

The chlorophylls are the essential components for photosynthesis and occur in chloroplasts as green pigments in all photosynthetic plant tissues. They are bound loosely to proteins but are readily extracted in organic solvents such as acetone or ether. Chemically each chlorophyll molecule contains a porphyrin (tetrapyrrole) nucleus with a chelated magnesium atom at the centre and a long chain hydrocarbon (phytol) side chain attached through a carboxylic acid group. There are at least five types of chlorophylls in plants. Chlorophylls a and b occur in higher plants, ferns and mosses. Chlorophylls c, d and e are only found in algae and in certain bacteria.

Chlorophyll is extracted in 80% acetone and the absorbance at 663nm and 645nm are read in a spectrophotometer. Using the absorption coefficients, the amount of chlorophyll is calculated.

Requirements

Reagents : Dilute analytical grade acetone to 80% acetone (prechilled), distilled water.

Glasswares: Volumetric flask, beakers, test tubes.

Equipments: Spectrophotometer, weighing balance, mortar and pestle, centrifuge, razor blade.

Procedure

- Weigh 1g of finely cut and well mixed representative sample of leaf or fruit tissue into a clean mortar and pestle.
- Grind the tissue to a fine pulp with the addition of 20ml of 80% acetone.
- Centrifuge (5000 rpm for 5 min) and transfer the supernatant to a 100 ml volumetric flask.
- Grind the residue with 20ml of 80% acetone, centrifuge and transfer the supernatant to the volumetric flask.
- Repeat this procedure until the residue is colourless. Wash the mortar and pestle thoroughly with 80% acetone and collect the clear washings in the volumetric flask.

- Make up the volume to 100ml with 80% acetone.
- Read the absorbance of the solution at 45, 663 and 652nm against the solvent (80% acetone) blank.

Calculation

Calculate the amount of chlorophyll present in the extract (mg chlorophyll per g tissue) using the following equations:

$$\text{mg Chlorophyll a/g tissue} = 12.7\,(A_{663}) - 2.69\,(A_{645}) \times \frac{V}{1000 \times W}$$

$$\text{mg Chlorophyll} = 22.9 \quad - 4.68\,(A_{645}) - 4.68\,(A_{663}) \times \frac{V}{1000 \times W}$$

(b/g tissue)

and

$$\text{mg total chlorophyll} = 20.2\,(A_{645}) + 8.02\,(A_{663}) \times \frac{V}{1000 \times W}$$

(per g tissue)

Where, A = Absorbance at specific wavelength

V = Final volume of chlorophyll extract in 80% acetone

and W = Fresh weight of the tissue extracted

Note :

The amount of tissue taken for extraction may be varied. Accordingly amount of 80% acetone used may be altered so that the final extract has a volume based on 10mg plant material in 1ml of acetone.

Observations

....

....

....

....

Experiment – 3 : To Separate Anthocynins by Paper and /or Thin layer Chromatography

Principle

Thin layer chromatography is based upon the principles of column and partition chromatography separation of a mixture in this cases is achieved over a thin layer of aluminium oxide or silica gel to which they are absorbed by different physical forces. The resolved mixture may be extracted with suitable solvents later on.

Requirement

Reagents : Rose petals, flowers of impatient *Basamina* L., 1% HCl (v/v) in 95% ethanol, Solvent I(n-butanol :acetic acid:water ::3:1:1), Solvent II(n-butanol: acetic acid : water :: 5:1:4), NH_4OH, $AlCl_3$ spray reagent (Dissolve 2.0 gm aluminum chloride ($AlCl_3$) and dilute to 100 ml with distilled water).

Glass/Plastic ware: test-tube, Petri dish, beakers, conical flask.

Equipments: automizer, evaporating device (fig 3.1) ultraviolet light source, thin-layer chromatography equipments.

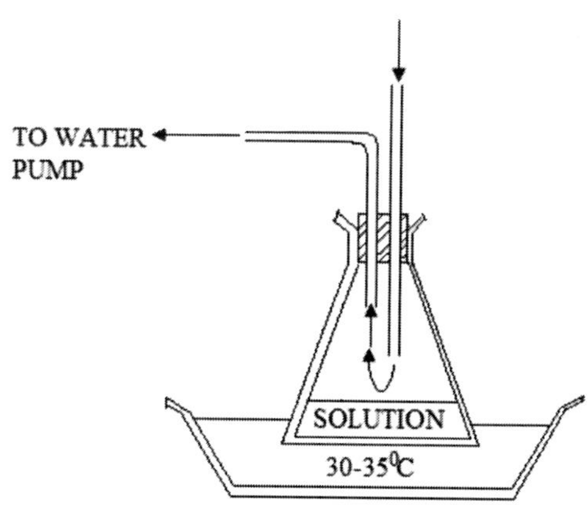

Fig 3.1. The evaporating device for lycopene extraction

Procedure

Extraction: Take a few buds and flowers of balsam belonging to the 3 stages of development as under. Stage 1: Buds closed, petal colour not developed. Stage2: Closed buds with well developed colour in sepals. Stage 3: Fully open flower with deeply coloured sepals. Homogenize the buds of stage1, stage 2 and the petals of the flower of stage 3 separately with 3 ml of 1% HCl in 95% ethanol, these may be kept in ice chamber or refrigerator for future use.

(a) **Paper chromatography (two dimensional):** Place a drop of about 6mm diameter of the homogenate in the usual way, on a chromatography paper. Use a new chromatography sheet for each stage of the bud. Place the papers in the chromatography chamber equipped with the solvent butanol: acetic acid: water in the ratio 3:1:1 (BAW) and allow it visible light and short wave ultraviolet light. Mark the spots with pencil and calculate the R_f value.

Run the chromatogram a second time with a change in the previous direction exactly by 90 in a jar containing 10 % acetic acid solution for about 2 hours. Remove the papers and dry them in air. Note the position of the spots in visible or ultraviolet light and mark the coloured fluorescent component by a pencil. Hold the paper over concentrated NH_4OH so that the spots become again under visible and ultraviolet light as before.

(b) **Thin-layer chromatography:** Use may also be made of thin layer chromatography the ratio of the solvent should be 5:1:4 (butanol: acetic acid: water).Trace the distinct spots that develop after drying. Mark under visible and ultraviolet light. Spot the component and calculate their R_f values.

Observations

....

....

....

....

Experiment – 4 : To Separate and Characterize the Anthocyanin Pigments

Principle

Anthocyanin are the most important and widespread groups of colouring matters in plants. These intensity coloured water-soluble pigments are responsible for nearly all colours in leaves, petals and fruits of higher plants. They are based chemically on a single aromatic structure that of cyanidin and all are derived from this pigment by addition or by subtraction of hydroxyl group or by methylation or by glycosylation. There are six common anthocyanidins (anthocyanin, aglycones formed when anthocyanins are hydrolyzed with acid) which differ in the nuture of the sugar (often glucose, but may also be galactose, rhamnose xylose or drabinose), the number of sugar units (mono di or tri glycosides) and the position of attachment of sugar (usually to the 3′ hydroxyl or to the 5′ hydroxyl). The alcohol extract of the sample of the sample is treated with HCl in aqueous methanol followed by anthocyanin reagent.

Requirements

Coloured flowers or red colored leaves and stems of plants.

Reagents : adsorposil-2 gypsum free cellulose MN-300, butanol, 2N HCl, acetone, formic acid, acetic acid (glacial), ethanolic $AlCl_3$, distilled water.

Glass/Plastic ware: Test-tube, petri dish, beakers, conical flask, glass plate for TLC, watch glass.

Equipments: Colorimeter, chromatography chamber, weighing balance, centrifuge, water bath, micropipette.

I-Procedure

- Homogenize 10 gm of different coloured plant materials separately in four volumes of 1% HCl in absolute methanol. Centrifuge the homogenate. Concentrate on water bath and store in a freeze away from light.
- The extract may be separated by thin layer chromatography (TLC). Dip two thin glass plates together in a slurry consisting of 30 mg silica acid (or cellulose) in 60 ml of distilled water for a couple of seconds. Take out and as the excess of the acid drains off dry the plates at room temperature for 2-4 hours

(when non-aqueous solution are to be used the plates should be activated by placing in a temperature of 120°C for 30 to 60 min).

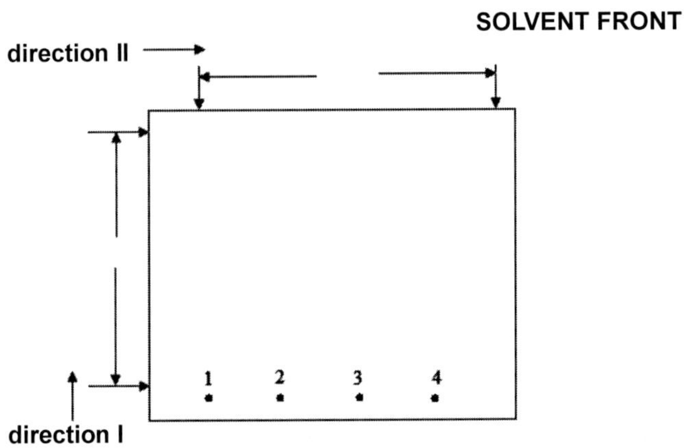

Chromatography paper showing the place for sports Fig.(1-4), and the directions in which the chromatogram is run

- Spots at a distance of at least 5cm from each other in one row at place shown (fig.) without disturbing the surface of the silica acid. Place the plate in a chamber saturated with acid containing BAW (n-butanol: glacial acetic acid: water in the proportion 5:1:4 mix, allow to settle and form two layers) in a manner that the lower edge dips in the solution. Allow the solvent to ascend 10-15 cm. Take the plate out and let it dry completely. Count the number of spots that have migrated from their original place and determine the R_f values. Observe these in day light as well as ultra-violet light.
- For two dimensional chromatography rotate the plate at 90^0 angle, place it in another chamber saturated with acid containing EFW (ethyl acetate: formic acid: water in the proportion of 8:2:3). Allow complete run in this solvent. After the development, allow to dry thoroughly, count the number of spots and determine R_f values.
- Remove the pigments from the plate by scraping the absorbent separately into a watch glass. Divide the material into two equal parts by weight. Dissolve one half in 5 ml 1 % HCl in methanol and the other half in ethanolic $AlCl_3$ (50gm of $AlCl_3$ mixed in ethanol and the volume made upto 1 litre. Shake well and allow salt to dissolve completely). Calculate the R_f value measure the colour, O.D. and % transmission of each spot.

- Identify the anthocyanins isolated by a reference to the R_f value.

  This method may be used to evaluate the effect of different light temperature, pH and photoperiod on the formation of anthocyanins.

II-Procedure

- Grind a known weight of fresh plant material in alcohol.
- Filter or centrifuge and collect the extract.
- Pipette 1ml of the alcohol extract into the test tube and add 3ml of HCl in aqueous alcohol.
- Add 1ml of anthocyanin reagents to the samples.
- Prepare the blank in the same manner by adding 1ml of methanol- HCl instead of anthocyanin reagent.
- After 15 min of incubation in the dark, measure the absorbance at 525 nm against the blank.
- Calculate the amount of anthocyanins present in the sample from a standard curve prepared with cyanin hydrochloride.

Notes :

1. *10μg of cyanin hydrochloride/ml in methanol-HCl = absorbance of 0.405 in a 1.0 cm cell at* A_{525}.
2. *Alternatively, the anthocyanin content may be expressed as* A_{525} *values.*

Observations

....

....

....

....

Experiment – 5 : Colour Development in Apples under the Influence of Quality of Light

Principle

Plant pigments may be conveniently classed under two groups viz. vacuolar pigments, and plastid pigments. The vacuolar pigments are water soluble pigments which occur in solution in the vacuolar sap. The colours of the flowers and beet root are examples of this class and include flavonoids, anthocyanins and betacyanins.

The pigments found inside the specialized masses of protoplasm lying in the cytoplasm are the plastid pigments. These may be colourless (leucoplasts), green photoactive (chloroplasts) and plastids of any colour other than white and green (chromoplasts).

The chloroplast pigments consist of

i) Chlorophylls (chl. a bluish green colour and chlorophyll b of olive green colour), and

ii) Carotenoids which contain carotenes (orange coloured) and xanthophylls (yellow coloured).

Chlorophyll a found in all green plants, is involved in the light reaction of photosynthesis. Chlorophyll b is found in green tissue of all higher plants, vascular and non- vascular cryptogams and in the chlorophyceae among algae. Chlorophyll b is absent in brown and red algae. Chlorophyll c and chlorophyll d may be present in the brown and red algae, respectively. Chlorophylls are insoluble in water but soluble in alcohol in alcohol, acetone and ether.

Flowers and fruits of different colours, other than white or green, have the colouring pigments which glycosidic compounds, the anthocyanin. The flavonols and flavones are water soluble include red and blue anthocyanins as well as the yellow flavonoids and flavones. Carotenes are also yellow but these are soluble in lipid and not water. The flavonoids are generally dissolved in the aqueous cell sap. Carotenes are localized in the plastids. They may impart yellow lemon, tangerine, red-orange, orange yellow colours. Both the types of pigments may be present in the same tissue (fig. 5.1).

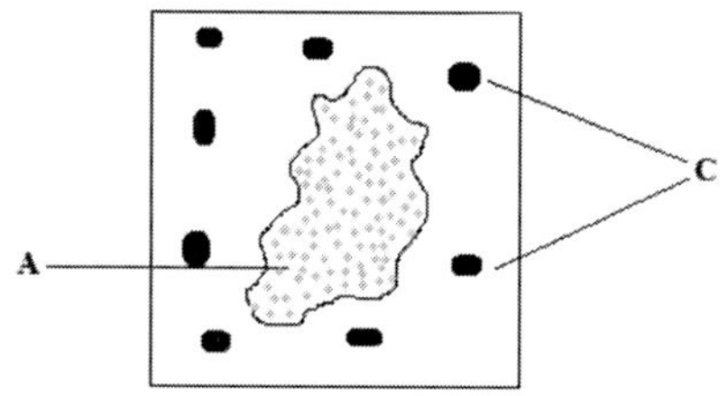

Fig. 5.1 : Location of Anthocyanins (A) and Carotenes (C) in plant cell.

Anthocyanins are widely distributed in higher plants but are rare in lower plants. Anthocyanin may aid as attractants of pollinators, sometimes they may also serve as inhibitor of bacterial growth pelargondin-3-monoglucoside or a means to classification by texonomists.

The sugar-free components of anthocyanin, the aglycon are called anthocyanidins. The anthocyanins sometimes change their colour due to change in pH of the cell sap. Cyanin, from rose petal, is red at pH of the cell sap. Cyanin, from rose petal, is red at pH 3 or less, violet at pH 8.5 and blue at pH 4.8.

Anthoxanthins are also flavonoid pigment. The glycon of these pigments are either flavonol on flavones itself. Both of them have got double bond ketonic oxygen at 4-carbon position but true flavones lack 3 hydroxyl group present in flavonol and anthocyanidins.

Requirements

Apples of green colour (apple variety must be capable of developing a red colour)

Reagents: 5% Clorox solution

Glass/Plastic wares: cellophane paper of different colours.

Equipments: artificial light source.

Procedure

Cover well washed and sterilized (with a 5% Clorox solution for 1 min.) one whole apple each in (1) black paper (2) colourless plastic (3) blue and red cellophane (far-red filter) (4) red filter (5) blue filter (6) black paper with punched holes in which cellophane of different colours viz. (green, orange, red, blue green and blue, orange and green, red and blue, clear etc. is fixed (fig 5.2).

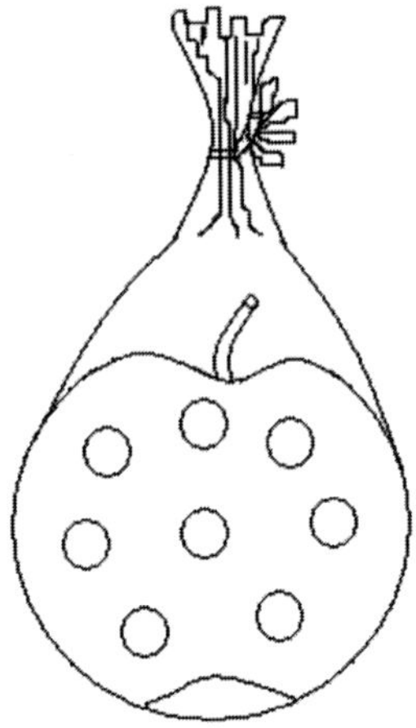

Fig. 5.2 : Apple in the covering of black peper having eight coloured windows. 1. green; 2. red; 3. orange; 4. blue; 5. red and blue; 6. green and blue; 7. orange and green; 8. clear.

Characteristics of Anthocynadins.

Name	Colour in Visible light	R_f value (paper chromatography) UV light	BAW	EFW
Apigeninidin	Orangey-yellow	Yellow-brown	0.74	0.44
Capensinnidin	purple	-	0.79	-
Cyanidin	Bluish- red or magenta	Pink	0.68	0.22
Delphindin	Purple-blue	Mauve	0.42	0.13
Flavone	-	-	-	-
Hirutidin	Bluish-red or magenta	Mauve	0.66	0.36
Luteolmidin	Orange	Red-brown	0.56	0.35
Malvidin	Purple	Mauve	0.58	0.27
Pelongonidin	Scarlet red	Orange-red	0.80	0.33
Peonidin	Bluish red or magenta	Pink	0.71	0.30
Petunidin	Purple blue	Mauve	0.52	0.20
Rosinidin	Bluish red or magenta	Pink	0.77	0.39
Tricetinidin	Orange	-	0.38	0.28

Place all the 6 apples under illumination for 3 to 4 days and observe where the colour developed and report as below.

Observation

Treatment	Observation
1. No light black cover	
2. Red light	
3. Blue light	
4. Ordinary light	
5. Far-red light	
6. Coloured windows	

Experiment – 6 : To Separate and Identify the Lycopene Present in Plant Material

Principle

Lycopene is responsible for the red colour of tomato and the fleshy part of water melon. It is a carotene having the formula $C_{40}H_{56}$ through it has no nutritional value, its contribution to the colour of tomato has a great role in consumer acceptability. The carotenoids in the sample are extracted in acetone and then taken up in the petroleum ether. Lycopene has absorption maxima at 473nm and 503nm. One mole of lycopene when dissolved in one litre light petroleum (40-60^{0}C) and measured in a spectrophotometer at 530 nm in 1cm light path gives an absorbance of 17.2 x 104. Therefore a concentration of 3.1206 µg lycopene/ml gives unit absorbance.

Requirements

Reagents: acetone, benzene, absolute ethyl alcohol, 95% ethanol, anhydrous Na_2SO_4 (Solid), anhydrous $CaCO_3$ (Solid), Al_2O_3. Tomatoes (ripe).

Glassware: test-tube, petri dish, beakers, conical flask, Buchner funnel, Erlenmeyer flask.

Equipments: column chromatography apparatus, spectrophoto-meter, mortar pestle, Whatman No.1 filter paper, refrigerator.

Procedure

Lycopene extraction: Homogenize a ripe tomato in a mortar. Add 100 ml. of acetone containing a pinch of $CaCO_3$ and 100 ml benzene and blend it again for a long period. Filter the homogenate with Whatman No.1 filter paper through a Buchner funnel. Repeat the filtration several times with a small quantity of acetone which should not exceed 50ml. Add 200 ml of distilled water to the filtrate and rotate gently to mix the water with the lower layer. Pour the filtrate in a 500ml separating funnel and discard the bottom light yellow. Wash the upper orange coloured benzene layer (do not shake) with 100 ml water several times till no trace of acetone is left. Take the benzene solution into and Erlenmeyer flask fitted with a stopper and tubing place it in a water bath (with a temperature of 30 to 35 °C), concentrate the solution, add small amount of anhydrous sodium sulphate and store in refrigerator.

Column chromatography: Fit up a device as shown in the figure. Any of the adsorbents *viz.* alumina, silica MgO and cellulose may be used to form the column. At the top of the column place a filter paper disc and cover it tightly with 2 to 3 cm of pre-baked $CaCO_3$. Place another filter paper disk on top of the whole column. Pour benzene to the column and apply suction to pack the column tighter and to remove pockets of air. When the entire column is wetted with benzene add the concentrated pigment solution through a pipette. Add enough solution to give a band on the alumina about 1 cm thick. Then pour anhydrous benzene and attach a separating funnel containing benzene on top of the column as shown in figure.

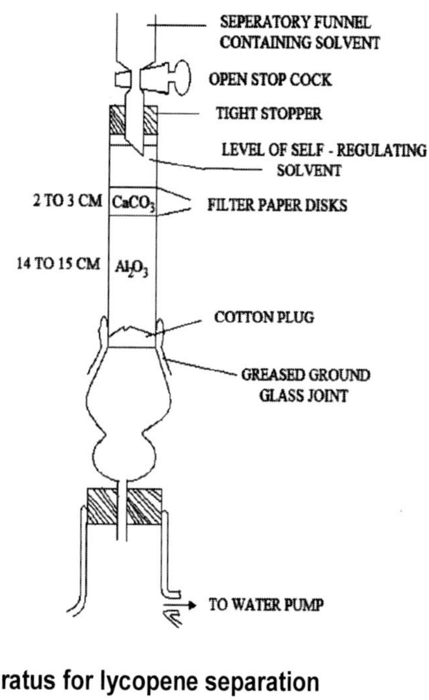

Apparatus for lycopene separation

Run anhydrous benzene through the column slowly, keep the top of the column covered with benzene throughout the experiment and continue the suction. Five bands (yellow band, orange-red band, thick red band, red band and brown band) are formed. First, band may be developing on the $CaCO_3$ column but later these are washed down and separated. Continue the process of development by using about 200 ml benzene for an hour. The yellow band of á and b carotene washes right in the flask. The brown band is at the top of alumina. Red is near the top, below which is the thick red and then the orange red. Add 2 drops of 100 % ethanol per 100 ml dry benzene to catalyze the process. After the separation of bands 2 and 3 allow the column to suck dry for 30 seconds. Separate out each of the alumina bands in a small beaker and cover with ethanol. Stir, through a piece of filter and collect the eluted pigment in small bottle. Add 20 to 30 ml of benzene to the ethanol elutes. Keep in a separating funnel and remove the alcohol by repeated washing with water. Dry the benzene solution with anhydrous sodium sulphate.

Absorption banding of lycopene: Test the absorption of the pigment solution against the solvent blank, 455, 470 and 500 nm are the three optical densities maximum for lycopene. Measure the absorption from 420 to 520 nm. Plot the optical density values on the ordinates against the respective wave lengths on the abscissa. Compare against the standard.

Observations

....

....

....

....

Experiment – 7 : To Isolate the Chloroplast Pigments on the Basis of their Differential Solubility in Various Organic Solvents and their Absorption Properties using Paper Chromatography

Principle

Besides the green colour, which is the predominant colour of the plant kingdom, higher plants also contain yellow pigments. The green colouring pigment of the plants is given the blanket them of chlorophyll but the yellow chloroplast pigments belong to the group of compounds known as carotenoids. Higher plants and green algae contain chlorophylls a and b only.

Carotenoids is a group of orange, red, yellow and brownish pigments. These pigments are classified into two different groups (a) carotenes which are generally orange, yellow, and (b) xanthophylls these are mostly yellow, brownish pigments. β - carotenes is most abundant in plants and is invariably present in the chloroplasts. Lutein is the most abundant xanthophylls in leaves. Other leaf xanthophylls include zeaxanthin, cryptoxanthin and xanthin etc.

The chloroplasts as well as the carotenoids are not soluble in water but these pigments are readily soluble in various organic solvents such as acetone, ether and alcohol etc. in solution chlorophyll is highly florescent emitting a deep colour when illuminated. In the living plants, most of this energy responsible for fluorescence is used in photosynthesis, and hence intact plants do not fluoresce appreciably.

Requirements

Reagents: Distilled water, Acetone, methyl alcoholic KOH, petroleum ether.

Glass/plastic wares: Petri dishes, Microscope slides, beakers, test tubes, buchner funnel, measuring cylinder.

Equipments : Analytical balance, buchner funnel, Centimeter rult graduated to 0.1 mm or a calibrated ocular micrometer to dissecting microscope, Forceps, razor blade, weighing balance.

Procedure

(A) **Extraction of leaf pigments and their separation by differential solubility:** Finely cut leaf material will be blended in acetone (30 gm in 100 ml of acetone) for 2 minutes, and then filtered on a buchner funnel. This extract contains both the water soluble and the chloroplast pigments.

Put 40 ml of petroleum ether into a separatory funnel and add 40 ml of the acetone extract. Now carefully add 60 ml water down the side of the funnel. Rotate the funnel until upper layer is quite green. Support the funnel by a ring and stand, allow the layers to separate and then draw off the lower acetone water layers. Discard this solution which contains the water soluble anthocynins and flavones. Wash the petroleum ether three times carefully by adding 50 ml water, gently rotating the funnel, and draining off the water.

Take 5 ml of petroleum ether solution into a test tube and observe it first in transmitted light then in reflected light under the ultraviolet lamp. Note in your report the colours observed. Return the solution to the separatory funnel.

To the clean petrol ether solution add 40 ml of 92% methanol, mix by rotation and then draw off methyl alcohol solution into a beaker. Chlorophyll a and carotenes are now in the petrol ether and chlorophyll and xanthophylls are in the methanol solution. Put the methanol solution into the separatory funnel, add 25 ml of ethyl ether and mix by rotating. Now add distilled water 25 ml at a time till two layers of liquid appear. Usually five or more additions are required. Discard the lower methyl alcohol layer.

Now put 15 ml each of the ethyl ether solution (containing chlorophyll b and xanthophyll) and (containing chlorophyll a and carotene) into a second tube. Carefully pour 10 ml of fresh 30% methyl alcoholic KOH down the wall of each test tube. Observe the brown layer which forms between the solvents. Shake the contents and observe for 10 min. Note the colour changes. Now add 15 ml of water to each tube, shake and permit the two layer of potassium salt of chlorophyll a or b to separate the case may be.

(B) **Separation of leaf pigments by paper chromatography:** Immerse about 1 gm fresh leaf tissue (tomato) in boiling water for a few minutes to kill the cells. Remove leaf material, blot off excess water with paper toweling, grind in a mortar with about1 mg fine grained quartz sand and several successive 10 - 15 ml portions of 50:1 petroleum ether: methyl alcohol solution. Grind very thoroughly until only a few milliliters of solvent remain.

Decant approximately 5 ml of this pigment into a test tube and insert a strip of filter paper cut to a slightly lesser diameter than the tube, chlorophyll b is absorbed more strongly than carotenes are not absorbed, and will collect at the top of the strip as the solvent evaporates. Mark positions of chlorophyll a and b identify the carotenoids and mount the chromatogram in your note book.

Observations

....

....

....

....

Experiment – 8 : Quantitative Determination and Study of Absorption Spectrum of Chlorophylls Extracted from the Leaves of Higher Plants

Principle

In this experiment you will determined the amount of total chlorophyll, chl a and chl b present in the leaf of the given plant material and determine the absorption spectrum of chlorophyll extracted from leaves.

Requirements

Reagents: Distilled water, 80% acetone.

Glass/plastic wares: Microscope slides, Petri dishes, beakers, test tubes, buchner funnel, measuring cylinder.

Equipments: Analytical balance, Centimeter rult graduated to 0.1 mm or a calibrated ocular micrometer to dissecting microscope, Forceps, weighing balance, whatman No. 1 filter paper, mortar and pestle, spectrophotometer.

Procedure

(A) **Chlorophyll Extraction:** The entire procedure to be carried out in dark room avoiding direct contact with light.

Place 1 gm (fresh wt.) of small pieces of leaf material provided to you, into a clean mortar. Add 40 ml of 80% acetone (v/v) and grind the tissue to a fine pulp (grind for about three minutes).

Carefully transfer the resulting green liquid to a buchner funnel containing a pad of whatman No. 1 filter paper. While filtering the extract (with suction) repeat the grinding of the pulp with a fresh 30 ml aliquot of 80% acetone. After 3 to 4 min. filter the second extract as before into the flask containing the first extract.

After the second extraction the tissue should be devoid of chlorophyll. If not, repeat the grinding with a fresh 20 ml aliquot of 80% acetone. Then, filter the slurry into the flask containing the other filtrates. With 10 ml of 80% acetone rinse, the mortar and sides of the funnel to ensure that all the chlorophyll is collected. For calculating the amount of chlorophyll present, adjust the final volume of the filtrate to 100 ml by adding 80% acetone.

If the plant material from which chlorophylls are to be extracted is less than 1 gm adjust the volume in such a way so that the final extract has a volume based on the ratio that 10 mg plant material extracted will come 1 ml of acetone.

(B) **Chlorophyll determinations:** Read and record the absorbance of the chlorophyll extract in 10 ml cell with a spectrophotometer set at 645, 663 and 652 nm. Be sure to read against an 80% acetone solvent blank. Then calculate the amount of chlorophyll present in the extract on the basis of milligrams of chlorophyll per gm of leaf tissue extracted according to the following equation:

$$\text{Mg chlorophyll a/g fr. Wt. } 12.7(A_{663}) - 2.69\,(A_{645}) \times \frac{V}{1000 \times W}$$

$$\text{Mg chlorophyll b/g fr. Wt. } 22.9(A_{645}) - 4.68\,(A_{663}) \times \frac{V}{1000 \times W}$$

$$\text{Total chlorophyll } 22.9(A_{645}) + 8.02\,(A_{663}) \times \frac{V}{1000 \times W}$$

In the above equations A represents the absorbance of chlorophyll extract at the specific wavelength, V the final volume of the solution and W gm of tissue extracted.

(C) **Absorption spectrum of chlorophyll in 80% acetone**: Dilutes a portion of the extract used above with sufficient 80% acetone to obtain transmittance reading to about 30 to 35%. Be sure to use an additional cell fitted with 80% acetone for reagent blank.

Take and record the absorbance reading of the diluted chlorophyll extract at 30 nm intervals over a range of 350 to 700 nm. At points of maximum absorption take additional readings at 5 nm intervals.

Calculate the amount of chlorophylls in milligrams per gram of leaf material extract. Also include a graph illustrating the absorption spectrum of chlorophyll in acetone with absorbance reading plotted on ordinate against the wave length of abscissa.

Observations

....

....

....

....

Experiment – 9 : Isolation of Active Chloroplasts from Leaves and Observation of Hill Reaction Activity of Chloroplasts

Principle

The experiment is designed to acquaint you with the techniques for the isolation of chloroplasts and to characterize the activity of chloroplasts preparation using Hill reaction activity. 2,4 - Dichlorophenol indophenol, a blue dye in its oxidized form, will be used as Hill oxidant. The bleaching of the dye (colourless form, produced) is the criterion for the presence of Hill activity.

Requirements

Reagents: 0.35 M NaCl, Tris - buffer pH 8.0 & 8.3, acetone (80%), phosphate buffer (pH 6.5, 0.1 M), 2.5 x 10^{-4}M, 2 - 4 Dichlorophenol indophenol dye. Fresh spinach (*Spinacea oleracea*) leaves

Glass/Plastic wares: Beakers, test tubes, measuring cylinder.

Equipments: Centrifuge, razor blade, refrigerator, ice bucket.

Procedure

Isolation of chloroplasts: Wash the leaves in cold distilled water and remove midribs from them. Homogenize 50 gm of sliced spinach leaf blades for 30 sec. in 100 ml 0.35M ice cold NaCl & 10 ml of ice cold 0.2M Tris buffer (pH, 8.0) at 4^0C. Filter the homogenate through a double layer of cheese cloth dampened with buffer in a beaker kept in ice bath centrifuge 30 ml of the green filtrate per minute at 300 x g auto remove unbroken cell and cell deodorized in a refrigerated centrifuge. Decant the supernatant in a cold tube and centrifuge it at 1000 x g for 10 min. to sediment the intact chloroplasts. Discard the supernatant. Place the tube containing chloroplasts pellet into an ice bucket. Suspend the sediment chloroplasts into 2 - 3 ml of 0.25M NaCl (cold by gentle stirring or shaking). Dilute the suspension to 30 ml with 0.35M cold NaCl solution. Repeat the centrifugation at 1000 x g for 10 min. discard the supernatant and resuspend the chloroplasts pellet in 10 ml of cold 0.35M NaCl. Store the chloroplast tube into ice bucket

Determination of chlorophyll content: Dilute 0.1 ml of the chloroplast preparation with 20 ml of 80% cold acetone. Filter the solution through Whatman No.1 filter paper. Place the solution in a cuvette and read the O.D. at 652 nm using 80% acetone as a blank.

Multiply O.D. with 5.8 to obtain chlorophyll concentration in mgs per ml of the suspension.

Dye reduction by chloroplasts: To each of the three clean, dry tubes add 2 ml 0.1 M phosphate buffer (pH 6.5) 6 ml cold distilled water. Note add the following: Tube 1: 2 ml of boiled chloroplasts. Tube 2: cover with carbon paper and then add 2 ml fresh chloroplasts.

Immediately after adding read O.D. of each tube at 650 nm quickly. Transfer tube 1 & 3 to bright light and tube 2 to darkness. Take O.D. readings at 5 min. intervals for about 30 min. or until the dye are maximally bleached.

Result

Plot O.D. values (vertical axis) against time (horizontal axis). Discuss differences between the shape of curves in the three tube and discuss. The 2,4 – Dichlorophenol indophenol dye is bleached as a result of the following oxidation – reduction reaction:

2,4 - Dichlorophenol indophenol + H_2O light dye - H_2 + SO_2 (reduced chloroplast) (colourless)

Experiment – 10 : To Extract the Curcumin Pigment

Principle

Curcumin is quantitatively extracted by refluxing the material in alcohol and is estimated spectrophotometrically at 425 nm.

Requirements

Reagents: Absolute Alcohol, Turmeric powder

Glass/Plastic wares: Stoppered Flask, beakers, measuring cylinder

Equipments: Air-condenser, weighing balance, heating mantle, spectrophotometer.

Procedure

- Dissolve 0.2-0.5 g of weighed, moisture free turmeric powder in 250 ml of absolute ethanol.
- Reflux the contents in the flask fitted with an air-condenser over a heating mantle for 3-5 h, compensate alcohol loss if any due to evaporation by adding alcohol freshly into the flask.
- Cool and decant the extract into a volumetric flask and make up the volume.
- Dilute a suitable aliquot (1-2 ml) to 10 ml with absolute alcohol. Measure the intensity of yellow colour at 425 nm in a spectrophotometer.

Calculation

Work out the Curcumin content using the formula

$$\text{Curcumin content g/100g} = \frac{0.0025 \times A_{425} \times \text{Volume made up} \times \text{Dilution factor} \times 100}{0.42 \times \text{Weight of the sample (g)} \times (1000)}$$

Since 0.42 absorbance at 425 nm = 0.0025 g Curcumin.

Note :

Alcohol is inflammable. Use no direct flame.

Observations

....

....

....

....

Additional Reading

Boyer R.F. (1993). Modern Experimental Biochemistry. 2nd ed. The Benjamin/Cummings Publishing Company, Inc. California. U.S.A.

Holme. D.J. and Peck H. (1998). Analytical Biochemistry. 3rd ed. Addision Wesely Longman Ltd. Essex. U.K.

Howcroft, D.M. (1997). Electrophoresis the Basics. IRL Press, Oxford U.K.

Wilson K. and Walker J. (1995). Principles and Techniques of Practical Biochemistry. 4th ed Cambridge U.K.

❑❑❑

Unit -9

Phenolics

Phenolics are broadly distributed in the plant kingdom and are the most abundant secondary metabolites of plants. Plant polyphenols have drawn increasing attention due to their potent antioxidant properties and their marked effects in the prevention of various oxidative stress associated diseases such as cancer. Polyphenols are a structural class of mainly natural, but also synthetic, and semisynthetic, organic chemicals characterized by the presence of large multiples of phenol structural units. The number and characteristics of these phenol structural underlie the unique physical, chemical, and biological (metabolic, toxic, therapeutic, etc.) properties of particular members of the class. The term polyphenol derives from poly-, from the ancient Greek word (polus, meaning "many, much") and the word phenol which refers to a chemical structure/substructure formed by attaching to an aromatic phenyl or benzenoid ring an alcohol-type hydroxyl (-OH) group (giving rise to the "-ol" suffix). The term "polyphenol" has been in use since 1894.

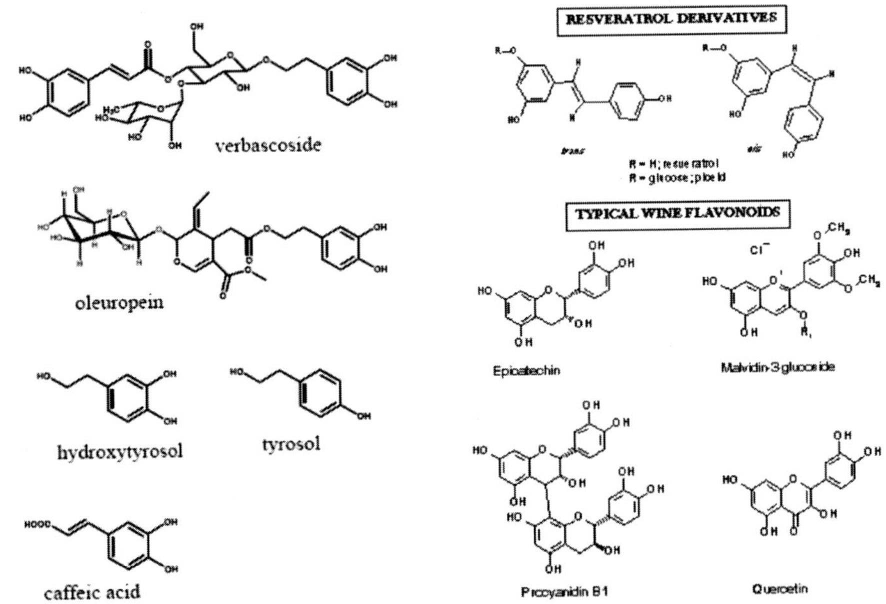

Fig. 9.1 : Structure of some phenolics

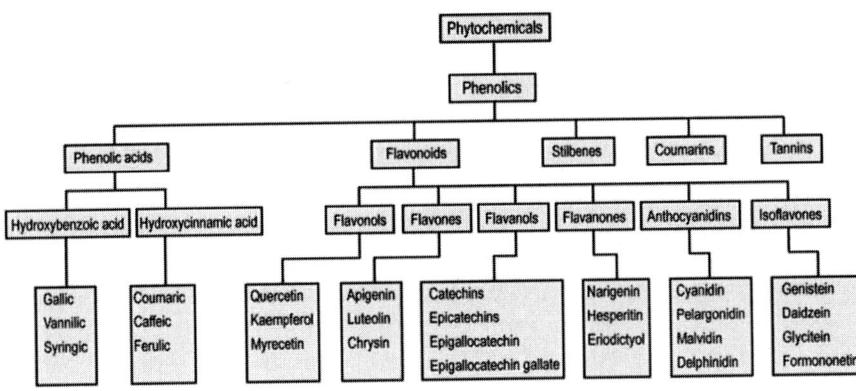

Fig. 9.2 : Typc of Phytochemicals

- *"Science is built up of facts, as a house is built of stones; but an accumulation of facts is no more a science than a heap of stones is a house."*

  Henri Poincare

- *"Failures do what is tension relieving while winners do what is goal achieving."*

  Dennis Waitley

Experiment – 1 : To Estimate Total Phenol Content in Plant

Principle

Phenols reduce phosphomolybdic acid and phosphotungustic acid alkaline condition to form blue colour complex which is measured colorimetrically at 750 nm.

Requirement

Reagents : Leaves, ethyl alcohol, whatman filter no.1.

Glass/Plastic wares: Sterile petri dishes, Sterile beakers, Sterile pasteur pipettes with rubber bulbs (wrapped in pipette box), Sterile disposable plastic syringes (1 ml, 10 ml), Funnel

Equipments: Spectrophotometer, Fine scalpel/razor blade, Fine forceps.

Procedure

- Plant samples infect with *R. solani* collect from the pots containing different levels of N,P,K,S,Zn and Fe and keep into a hot air oven to dry at 55 ± 5°C for 20 hrs.
- Dry samples and cut into small pieces which mix (< 1 mm) and keep into desiccators.
- Subsequently 1 g of the mix plant sample grind in a mortar with a pestle along with 20 ml of 80% ethyl alcohol.
- After grinding of sample mix $NaCO_3$ with the help of whatman filter paper No.1.
- One ml filtrate, 1 ml phenol reagent and 2 ml of $NaCO_3$ solution take together and volume adjust to 50 ml by adding distilled water.
- 20 ml of this preparation keep in a test tube for 1 hr and its optical density (O.D.) record with the help of a spectrophotometer at 750 nm against blank reagent. Simultaneously a standard curve prepare by using various concentration of gallic acid and concentration of phenols calculate in the test sample and expresse as phenols weight in mg per 100 g dried plant material.
- Quantity of phenols calculate with the help of following formula

$$\text{Total phenol (mg/100g dry wt. basis)} = \frac{\text{Amount of gallic acid from std. X Sample O.D. X volume made}}{\text{O.D. of known gallic acid} \times \text{Sample wt.} \times \text{Volume of aliquot}}$$

Observations

....

....

....

....

Experiment – 2 : To Estimate the Chlorogenic Acid

Principle

Chlorogenic acid is extracted with alcohol, dried and dissolved in acetone. It is reacted with titanium ion to form a coloured complex which is measured at 450 nm.

Requirement

Reagents: Titanium Reagent: 20% $TiCl_4$ in conc. HCl, Standard 25-200 ìg/ml Chlorogenic Acid in Acetone, Acetone, 80% Ethanol, 2.5 N HCl

Glass/Plastic wares: Beakers, test tubes, measuring cylinder, volumetric flask.

Equipments: Spectrophotofluorimeter, centrifuge, Micropipettes, pestle mortar, refrigerator.

Procedure

- Reflux twice a known quantity of defatted sunflower meal in 80% ethanol (adjusted to pH 4.0 with 2.5 N HCl) for 30 min (125 ml to 1 g meal).
- Discard the precipitate and collect 250 ml of the extract.
- Remove 0.5 ml samples and dry in a vacuum oven at 50^0C and 700 mm pressure for 2 h.
- Dissolve dried extract in 4.75 ml of acetone.
- Add 0.25 ml of $TiCl_4$.
- Read the colour at 450 nm against a reagent blank (acetone plus $TiCl_4$).
- Similarly treat the standards with $TiCl_4$ and read the colour intensity.
- Draw a standard curve and find out the chlorogenic acid content in the sample.

Calculation

Express chlorogenic acid content as g per 100 g sample.

Notes :

1. *If titanium chloride is not available, this estimation can be performed with Folin-Ciocalteau reagent. Dissolve the dried extract in 5 ml ethanol and proceed for Folin-Ciocalteau method (Refer phenol estimation). Similarly, prepare the standard in 80% ethanol and develop the colour with Folin-Ciocalteau reagent. In this method, there is a possibility of interference of tyrosine and phenylalanine.*
2. *The titanium chloride method can be followed for other phenols, too. But the wavelength of maximum absorbance differs. Total phenols can be measured at 410 nm, catechol and catechin at 430 nm.*

Observations

....

....

....

....

Experiment – 3 : To Estimate the Tannins by Folin-Denis and Vanillin Hydrocloride Methods

i) Folin-Denis Method

Principle

Tannin-like compounds reduce phosphotungstomolybdic acid in alkaline solution to produce a highly coloured blue solution, the intensity of which is proportional to the amount of tannins. The intensity is measured in a spectrophotometer at 700 nm.

Requirement

Reagents

- Folin-Denis Reagent: Dissolve 100 g sodium tungstate and 20 g phosphomolybdic acid in 750 ml distilled water in a suitable flask and add 50 ml phosphoric acid. Reflux the mixture for 2 h and make up to one liter with water. Protect the reagent from the exposure to light.
- Sodium carbonate Solution: Dissolve 350 g sodium carbonate in one liter of water at 70 – 80°C. filter through glasswool after allowing it to stand overnight.
- Standard Tannic Acid Solution: Dissolve 100 mg tannic in 100 ml of distilled water.
- Working Standard Solution: Dilute 5 ml of the stock solution to 100 ml with distilled water. One ml contains 50 mg tannic acid.

Glass/Plastic wares: Beakers, test tubes, measuring cylinder, volumetric flask.

Equipments: Spectrophotofluorimeter, centrifuge, Micropipettes, pestle mortar, refrigerator

Procedure

- Extraction of Tannin: Weigh 0.5 g of the powdered material and transfer to a 250 ml conical flask. Add 75 ml water. Heat the flask gently and boil 30 min. Centrifuge at 2,000 rpm for 20 min and collect the supernatant in 100 ml volumetric flask and make up the volume.
- Transfer 1 ml of the sample extract to a 100 ml volumetric flask containing 75 ml water.

- Add 5 ml of Folin-Denis reagent, 10 ml of sodium carbonate solution and dilute to 100 ml with water.
- Shake well. Read the absorbance at 700 nm after 30 min.
- If absorbance is greater than 0.7, make a 1+4 dilution of the sample.
- Prepare a blank with water instead of the sample.
- Prepare a standard graph by using 0 – 100 μg tannic acid.

Calculation

Calculate the tannin content of the samples as tannic acid equivalents from the standard graph.

ii) Vanillin Hydrochloride Method

Principle

The vanillin reagent will react with any phenol that has an unsubstituted resorcinol or phloroglucinol nucleus and forms a coloured substituted product which is measured at 500 nm.

Requirement

Reagents

- Vanillin Hydrochloride Reagent: Mix equal volumes of 8% hydrochloric acid in methanol and 4% vanillin in methanol. The solutions must be mixed just before use, and avoid using even if it is slightly coloured.
- Catechin-Stock Standard Solution: Prepare a standard solution containing 1 mg catechin/ ml methanol.
- Working standard: Dilute the above stock solution ten times. 10-100 ml (100 mg/ml).
- Preparation of the Extract: Extract 1 g of ground seed in 50 ml methanol. Mix occasionally by swirling. After 20-28 h, centrifuge and collect the supernatant.

Glass/Plastic wares: Beakers, test tubes, measuring cylinder, volumetric flask.

Equipments: Spectrophotofluorimeter, centrifuge, Micropipettes, pestle mortar, refrigerator.

Procedure

- Pipette out 1 ml of the supernatant.
- Quickly add 5 ml of vanillin hydrochloric reagent.
- Read in a spectrophotometer at 500 nm after 20 min.
- Prepare a blank with vanillin hydrochloric reagent alone.
- Prepare a standard graph with 20 – 100 μg catechin using the diluted stock solution.

Calculation

From the standard graph, calculate the amount of catechin, i.e., tannin in the sample as per the absorbance value and express the results as catechin equivalents.

Observations

....

....

....

....

Experiment – 4 : To Estimate the Lignins

Principle

Refluxing the sample material with acid detergent solution removes the water-solubles and materials other than the fibrous component. The left-out material is weighed after filtration, dried, treated with 72% H_2SO_4 and filtered, dried and ashed. The loss of weight on ignition gives the acid detergent lignin.

Requirement

Reagents: Acid Detergent Solution: Dissolve 20 g of cetyl trimethyl ammonium bromide in one liter of 1 N sulphuric acid, 72% H_2SO_4 (w/v), Acetone

Glass/Plastic wares: Beakers, test tubes, measuring cylinder, Round Bottom Flask and Refluxing Set, Sintered Glass Crucible- G2

Equipments: Spectrophotofluorimeter, Micropipettes, pestle mortar refrigerator, Muffle Furnace.

Procedure

A. Acid Detergent Fibre (ADF)

1. Place 1 g of powdered sample in a round bottom flask and 100 ml of acid detergent solution. Heat to boil in 5-10 min. Reduce heat to avoid foaming as boiling begins. Reflux for 1 h after the onset of boiling. Adjust boiling to slow, even level
2. Remove container, swirl and filter the contents through a preweighed sintered glass crucible (G-2) by suction and wash with hot water twice.
3. Wash with acetone and break up the lumps. Repeat acetone washing until the filtrate is colourless.
4. Dry at 100°C for overnight.
5. Weigh after cooling in a desiccator.
6. Express ADF content in percentage i.e., W/S x 100, where W is the weight of the fibre and S is the weight of the sample.

B. Determination of Acid Detergent Lignin (ADL)

1. Transfer ADF to a 100 ml beaker with 25-50 ml of 72% sulphuric acid. Add 1 g asbestos. Allow it to stand for 3 h with intermittent stirring with a glass rod.
2. Dilute the acid with distilled water and filter with preweighed Whatman No. 1 filter paper. Wash the glass rod and the residue several times to get rid of the acid.
3. Dry the filter paper at 100°C and weigh after cooling in a desiccator.
4. Transfer the filter paper to a preweighed silica crucible and ash the filter paper with the content in a muffle furnace at 550°C for about 3 h.
5. Cool the crucible in a desiccator and weigh. Calculate the ash content.
6. For blank take 1 g asbestos, add 72% H_2SO_4 and follow the steps 2-5.

Calculation

$$\% \text{ADL} = \frac{\dfrac{\text{Weight 72\% of } H_2SO_4 \text{ washed fibre}}{\text{(Test - Asbestos blank)}} - \dfrac{\text{Ash}}{\text{(Test - Asbestos blank)}} \times 100}{\text{Weight of sample}}$$

Notes :

1. *NDF = Hemicellulose + Cellulose + Lignin + Minerals.*

   ADF = Cellulose + Lignin + Minerals.

   Hemicelluloses = NDF – ADF.

   Cellulose = ADF – Residue after extraction with 72% H_2SO_4

   Lignin = Residue after extraction with 72% H_2SO_4 *– Ash.*
2. *Wet materials can also be used for lignin estimation, provided the wet sample is equivalent to 1 g of dry material.*
3. *If the weight loss of asbestos blank on ashing is below 0.002 g/g, discontinue the determination of blank.*

Observations

....

....

....

....

Experiment – 5 : To Estimate the Capsaicin by Colorimetric and Spectrophotometric Methods

i) Colorimetric Method

Principle

The phenolic group in capsaicin reduces phosphomolybdic acid to lower acids of molybdenum. The resulting component is blue in colour and is read at 650 nm. The colour intensity is directly proportional to the concentration of capsaicin.

Requirement

Reagents

- 0.4% Sodium Hydroxide
- 3% Phosphomolybdic Acid
- Dry Acetone (Add about 25 g anhydrous sodium sulphate to 500 ml acetone of analytical grade at least one day before use).
- Stock Standard Capsaicin Solution: Dissolve exactly 50 mg capsaicin in 50 ml of 0.4% sodium hydroxide solution (1,000 mg/ml).
- Working Standard: Dilute 10 ml of the stock standard to 50 ml with 0.4% sodium hydroxide solution (200 mg/ml).

Glass/Plastic wares: Beakers, test tubes, measuring cylinder, volumetric flask.

Equipments: Spectrophotometer, centrifuge, Micropipettes, pestle mortar, refrigerator.

Procedure

1. Weigh 0.5 g dry chilli powder into a glass-stoppered test tube or volumetric flask.
2. Pipette out 10 ml dry acetone into the flask and shake it for 3 h in a mechanical shaker. Let the contents settle down or centrifuge (10,000 rpm for 10 min).
3. Pipette out 1 ml of the clear supernatant into a test tube and evaporate to dryness in a hot water-bath.
4. Dissolve the residue in 5 ml of 0.4% sodium hydroxide solution.
5. Add 3 ml of 3% phosphomolybdic acid.

6. Shake the contents and let stand for 1 h.
7. Filter the solution quickly into centrifuge tubes to remove any floating debris.
8. Centrifuge at about 5,000 rpm for 10-15 min.
9. Transfer the clear blue coloured solution directly into the cuvette and read the absorbance at 650 nm.
10. Run a reagent blank along with the test samples.
11. Prepare a standard using 0-200 µg capsaicin simultaneously i.e. pipette out 0.2, 0.4, 0.6, 0.8 and 1 ml of working standard solution and proceed as above.

ii) Spectrophotometric Method

Principle

Capsaicin is extracted from red pepper (chilies) with ethyl acetate and made to react with ethyl acetate solution of vanadium oxychloride. Then it is read at 720 nm. This method is sensitive and is useful to measure small quantities (less than 0.05%).

Requirements

Reagents

- Vanadium oxychloride (0.5%) in ethyl acetate.
- Pure capsaicin (0.01%) in ethyl acetate (10 mg in 100 ml).

Glass/Plastic wares: Beakers, test tubes, measuring cylinder, volumetric flask.

Equipments: Spectrophotometer, Micropipettes, pestle mortar, refrigerator.

Procedurc

- Grind the sample well to pass through No. 40 sieve.
- Place 2 g sample in a 100 ml volumetric flask.
- Let it stand for 24 h to extract (otherwise reflux the contents for 2.5 h) and then make up to volume.
- Dilute 1 ml of extract to 5 ml with ethyl acetate.
- Add 0.5 ml Vanadium oxychloride solution (just before reading) and shake.

6. Read at 720 nm in a spectrophotometer.
7. Subtract the absorbance value.
8. Prepare a standard curve using 0.5, 1.0, 1.5, 2.0 and 2.5 ml of standard capsaicin solution containing 50, 100, 150, 200 and 250 μg capsaicin respectively.

Calculation

$$\% \text{ capsaicin} = \frac{\mu\text{g capsaicin}}{1000 \times 1000} \times \frac{100}{1} \times \frac{100}{2}$$

$$= \frac{\mu\text{g capsaicin}}{200}$$

Observations

....

....

....

....

Additional Reading

Bergmeyer, H.U. (1983). Methods of enzymatic Analysis vol III and IV 3rd ed. Verlag Chemie Weihheim, Germany.

Coultate, T.P. (1996). Food-The chemistry of its components 3rd ed. Royal Society of Chemistry. London, U.K.

Day R.A. Jr and Underwood, AL (1993). Quantitative Analysis 6th ed Prentice - Hall of India, New Delhi.

Hector, T. and James, A.M. (1986). Assessment and Control of Biochemical Methods. John Wiley and Sons, Chichester, U.K.

❑❑❑

Unit -10

Anti Nutritional Factors

An **anti-nutritional factor** is a substance which, when present in human or animal foods, reduces growth. Examples are phytate, protease inhibitors (notably soybean trypsin inhibitor) and excessive dietary fiber.

Anti-Nutritional Factors

- **Protease inhibitors** inhibit the activity of trypsin, chemotropism and other proteases. They are found in legumes such as beans and peas, but also in cereals, potatoes, and other products. Their presence results in impaired growth and poor food utilization.
- **Amylase inhibitors** have a similar activity against amylases. Amylases are important in breaking down the structure of carbohydrates; they hydrolyze sugar and starches. Ultimately improving digestion of carbohydrates.

Fig. 10.1 : Anti Nutritional Food

- **Lectins** or humagglutinins are glycoprotein mainly found in legumes: beans, peas, lentils. Their presence results in poor food utilization and impaired growth.
- **Glucosinolates** are found in cabbage and related species. Effects upon the thyroid function have been demonstrated.

Fig. 10.2 : Anti Nutritional Products

- **Saponins** are found in soybeans, peanuts, sugar beets and others. Toxic effects have been shown.
- **Gossypol** is particularly important in cottonseed. Several toxic effects have been demonstrated.
- **Phytic acid** occurs in several vegetable products. Its presence may affect bioavailability of minerals.

Toxins

- **Cyanogens** are found in cassava, linseed, peas, beans and other products. They may cause food poisoning.
- **Mycotoxins** are very divers in chemical structure and in the characteristics of the toxins they produce. These toxins include the aflatoxins, the tricothecenes, the fumonisins, zearalenone, moniliformin and fusaric acid
- **Phytotoxins** (solanine, etc).

- *"Nobody climbs mountains for scientific reasons. Science is used to raise money for the expeditions, but you really climb for the hell of it."*

Edmund Hillary

Experiment – 1 : Estimation of Trypsin Inhibitor

Principle

The trypsin inhibitor activity is measured indirectly by inhibiting the activity of trypsin. A synthetic substrate (BAPNA) is subjected to hydrolysis by trypsin to produce yellow coloured p-nitroanilide. The degree of inhibition by the extract of the yellow colour production is measured at 410 nm.

Requirements

Reagents

- 30% Glacial Acetic Acid (v/v)
- Trypsin: Dissolve 6.25 mg lyophilized trypsin and make up to 25 ml with 0.01 M HCl. Dilute 2 ml of this solution to 25 ml for assay.
- Substrate: Benzoyl-DL-Arginine-paranitroanilide (BAPNA). Completely dissolve 40 mg BAPNA in 0.5 ml of dimethyl sulphoxide and then make up to 100 ml with Tris-HCl buffer pH 8.2.
- Tris-HCl Buffer pH 8.2: Weigh 6.05 g Tris (hydroxymethyl aminomethane) and 2.94 g $CaCl_2.H_2O$, dissolve in 900 ml water, adjust pH 8.2 with dil HCl and make up to 1000 ml with distilled water.
- Source of TI: Extract 0.5 g sample in 25 ml water by grinding in a pre-chilled mortar and pestle. Extract the ground sample in a refrigerator for 2-3 h with occasional shaking for complete extraction of TI. Centrifuge the homogenate at 12,000 rpm for 20 min at 4-6°C. dilute 1 ml of the supernatant to 10 ml with distilled water and use as TI source.

Glass/Plastic wares: beakers, test tubes, measuring cylinder.

Equipments: weighing balance, micropipette, spectrophotometer, water bath.

Procedure

- Pipette out 0-1 ml of extract in duplicate sets of test tubes, one to serve as endogenous (E) and the other test (T).
- Make up the volume to 2 ml with buffer in the endogenous set.

- Make up the volume to 1 ml in the test set.
- Add 1 ml of trypsin solution (20 μg) to each tube in the test set. Pipette out into a separate test tube 1 ml of buffer and 1 ml of trypsin solution for standard (S).
- Incubate all the tubes in a water bath at 37°C.
- After a few minutes, add 2.5 ml of substrate (1 mg BAPNA) to each tube.
- Allow the reaction to proceed for 10-60 min at 37°C.
- Stop the reaction by adding 0.5 ml of 30% glacial acetic acid.
- Read the absorbance at 410 nm in a spectrophotometer.
- Determine the protein content in the extract by the method of Lowry et al.

Calculation

Find out T-S absorbance.

Plot the absorbance against the volume of extract.

Determine the aliquot size of the extract required to inhibit 50% of the trypsin activity (S/2). That aliquot size is considered to be one unit of trypsin inhibitor.

One unit of activity corresponds to that amount of trypsin inhibitor in μg protein which gives 50% inhibition of enzyme activity under experimental conditions. The trypsin inhibitor activity is expressed as trypsin inhibitor units (TIU) per gram sample or per mg protein. (The dilutions of trypsin inhibitor source were made in such a way that 0.5 ml produces 50% inhibition).

Observations

....

....

....

....

Experiment – 2 : Determinaton of Cyanogens

Principle

Hydrocyanic acid which is evolved from the sample, forms a red coloured compound with sodium picrate and the intensity is measured at 525 nm.

Requirements

Reagents : Chloroform, Alkaline picrate solution: Dissolve 25 g sodium carbonate and 5 g picric acid in one liter of water, Standard hydrogen cyanide solution: Dissolve 0.241 g of KCN/liter of water. This gives a solution containing 100 µg hydrogen cyanide/ml.

Glass/Plastic wares: beakers, test tubes, measuring cylinder, conical flask.

Equipments: Whatman No. 1 filter paper: Cut filter paper into strips 10-12 cm long and 0.5 cm wide, and saturate them with alkaline picrate solution, spectrophotometer,

Procedure

- Homogenize 1 g of the sample in 25 ml water with 3-4 drops of chloroform.
- Place the homogenate in 500 ml conical flask.
- Place the saturated filter paper in the hanging position with the help of a cork stopper inside the conical flask.
- Incubate the mixture at room temperature (20°C) for 20-24 h.
- The sodium picrate present in the filter paper is reduced to reddish compound in proportion to the amount of hydrocyanic acid evolved.
- Elute the colour by placing the paper in a clean test tube containing 10 ml distilled water and compare it with standards at 625 nm.
- Preparation of the standard curve: (KCN is poisonous. Use propipettes, automatic pipettes or burettes for transfer of cyanide solution). Place 5 ml of the alkaline picrate solution and 5 ml of the KCN solution in a test tube. Heat for 5 min in boiling water. Deliver the following volumes from the above KCN alkaline picrate solution to six tubes: 0.1, 0.2, 0.4, 0.6, 0.8

and 1 ml. bring the volume of each test tube to 10 ml by adding distilled water. Press rubber stopper to the tubes and keep them in a cool place. Measure the absorbance at 525 nm. The amount of hydrogen cyanide in the above tubes is equal to 5, 10, 20, 30, 40 and 50 μg respectively. Prepare a blank with 10 ml distilled water.

Calculation

Calculate the hydrogen cyanide content of the sample from the standard graph.

Observations

....

....

....

....

Experiment – 3 : Estimation of Phytic Acid in Food Grains

Principle

The phytate is extracted with trichloroacetic acid and precipitated as ferric salt. The iron content of the precipitate is determined colorimetrically and the phytate phosphorous content calculated from this value assuming a constant 4 Fe: 6 P molecular ratio in the precipitate.

Requirements

Reagents: 3% Trichloroacetic acid, 3% Sodium sulphate in 3% TCA, 1.5 N NaOH, 3.2 N HNO_3, $FeCl_3$ Solution (Dissolve 583 mg $FeCl_3$ in 100 ml of 3% TCA), 1.5 M Potassium thiocynate (KSCN) (Dissolve 29.15 g in 200 ml water), Standard $Fe(NO_3)_3$ solution.

Glass/Plastic wares: beakers, test tubes, conical flask, measuring cylinder, Erlenmeyer flask.

Equipments: centrifuge, Whatman No.2 filter paper, water bath, centrifuge.

Procedure

- Weigh a finely ground (40 mesh) sample estimated to contain 5-30 mg phytate P into a 125 ml Erlenmeyer flask.
- Extract in 50 ml 3% TCA for 30 min with mechanical shaking or with occasional swirling by hand for 45 min.
- Centrifuge the suspension and transfer a 10 ml aliquot of the supernatant to a 40 ml conical centrifuge tube.
- Add 4 ml of $FeCl_3$ solution to the aliquot by blowing rapidly from the pipette.
- Heat the contents in a boiling water-bath for 45 min. If the supernatant is not clear after 30 min, add one or two drops of 3%sodium sulphate in 3% TCA and continue heating.
- Centrifuge (10-15 min) and carefully decant the clear supernatant.
- Wash the precipitate twice by dispersing well in 20-25 ml 3% TCA, heat in boiling water for 5-10 min and centrifuge.
- Repeat washing with water.
- Disperse the precipitate in a few ml of water and add 3 ml 1.5 N NaOH with mixing.

- Bring volume to approximately 30 ml water and heat in boiling water for 30 min.
- Filter hot (quantitatively) through a moderately retentive paper Whatman No. 2.
- Wash the precipitate with 60-70 ml hot water and discard the filtrate.
- Dissolve the precipitate from the paper with 40 ml hot 3.2 N HNO_3 into a 100 ml volumetric flask.
- Wash paper with several portions of water, collecting the washings in the same flask.
- Cool flask and contents to room temperature and dilute to volume with water.
- Transfer a 5 ml aliquot to another 100 ml volumetric flask and dilute to approximately 70 ml.
- Add 20 ml of 1.5 M KSCN dilute to volume, and read colour immediately (within 1 min) at 480 nm.
- Run a reagent blank with each set of samples.

Standard

Weigh accurately 433 mg Fe $(NO_3)_3$ and dissolve in 100 ml distilled water in a volumetric flask. Dilute 2.5 ml of this stock standard and make up to 250 ml in a volumetric flask. Pipette out 2.5, 5, 10, 15 and 20 ml of this working standard into a series of 100 ml volumetric flasks and proceed from set 16.

Calculation

Find out the μg iron present in the test from the standard curve, and calculate the phytate P as per the equation.

$$\text{Phytate P mg/100 gm sample} = \frac{\mu g\ \text{Fe} \times 15}{\text{Weight of sample(g)}}$$

Observations

....

....

....

....

Experiment – 4 : To Study the Cysteine Proteinase Inhibitor Assay

Principle

An artificial substrate, Benzoyl-DL-arginine p-nitroanilide (BAPNA) is hydrolysed by cysteine proteinases, similar to papain. The colour intensity of the released p-nitroaniline is measured at 410 nm. The colour intensity of the released p-nitroaniline in the sample may be compared with the standard p-nitroaniline or calculated using the molar extinction coefficient of p-nitroaniline.

Requirements

Regents

- Extraction Buffer: Sodium phosphate, 0.025 M (pH 7.0) containing 0.15 M NaCl
- Assay Buffer: Sodium phosphate buffer 0.25 M, pH 6.0 containing 2.5 nM EDTA Na_2^+ and 25 mM 2-mercaptoethanol
- Papain Solution: 25 mg/ml dissolved in assay buffer
- Benzoyl-DL-arginine-p-nitroanilide (BAPNA) 10 mM: Dissolve 108 mg of BAPNA in 6 ml dimethyl sulphoxide and make up the volume to 25 ml using assay buffer.
- Glacial acetic acid, 30% (v/v)
- Enzyme extract: The method described by Abe et al. (1987) was adopted for the extraction of cysteine proteinase inhibitor (CPI) with some modifications.

  Fresh source material (for example, Jackfruit seeds) were cut into small pieces using a razor blade, soaked overnight in the extraction buffer at 4°C and homogenized mechanically in a mixer grinder or 10 g dry seed powder/fresh leaves are extracted in 20 ml of ice-cold extraction buffer by grinding in a mortar and pestle. The homogenate was centrifuged at 10,000 g at 4°C for 15 min. the supernatant was heated at 65°C for 40 min. it was again centrifuged at 10,000 g for 20 min at 4°C to obtain the supernatant as crude enzyme extract and used.

Glass/Plastic wares: Beakers, test tubes, measuring cylinder.

Equipments: Spectrophotofluorimeter, Micropipettes, incubator, centrifuge.

Procedure

- Incubate papain (25 μL) and 0.5 ml of inhibitor extract together with 0.5 ml of assay buffer at 37°C for 20 min.
- Set a blank without inhibitor extract and another one without enzyme, replacing the enzyme or inhibitor with an equal volume of extraction buffer.
- Start the reaction by the addition of 200 μl of 10 mM BAPNA and incubate at 37°C for 45 min.
- Arrest the reaction by the addition of 200 μl of 30% acetic acid.
- Measure the absorbance of the solution at 410 nM.
- Draw a standard graph plotting absorbance vs. p-nitroaniline (μg).
- Calculate the amount of p-nitro aniline (μg) released in the samples and work out the % inhibition.

Units of Activity

One unit of papain activity is defined as the amount of enzyme that liberates one n mole of p-nitroaniline per minute under the assay conditions.

One unit of the inhibitor activity is defined as the amount of protein of the inhibitor extract that inhibits 1 unit of papain activity.

Observations

....

....

....

....

Experiment – 5 : To Study the Antifungal Activity of Cystatin Present in the Given Sample

Principle

The growth of one or more reference fungus is compared in the presence and absence of the cysteine proteinase inhibitory level is calculated.

Requirement

Reagents

- Crude extracts/partially purified fractions/purified inhibitor (for example, jackfruit cystatin)
- Potato dextrose agar medium:

  Peeled potato - 250 g

  Dextrose - 20 g

  Agar-15g

  Distilled Water-1L

  Adjust to pH 6.5
- Fungal cultures: (For example, *Alternarai alternate, Macrophomina phaseolina, Rhizoctonia solani and Pythium aphanidermatum*)
- Streptomycin sulphate (stock): Dissolve 1g/100 mL distilled water, and add 0.3-mL/100 mL medium.

Glass/Plastic wares: Beakers, test tubes, measuring cylinder.

Equipments: Centrifuge, Micropipettes.

Procedure

- Sterilize the potato-dextrose-agar-medium in an autoclave at 15 lbs pressure (1.05 kg/cm^2) for 20 min.
- In corporate streptomycin sulphate into the medium as mentioned above, prior to plating in order to avoid bacterial contamination.
- Add crude inhibitor extract or purified fraction of cystatin to the medium at a concentration of 600 µg/20 mL of the medium.
- Place the fungus culture at the centre of the petri plate, seal and incubate at room temperature.

- Set a control for each culture with no addition of inhibitor extract.
- Incubate the plates at room temperature for 2-3 days until the fungal growth occupies the entire plate (9 cm diameter) in the control plate.
- Measure the diameter (cm) of growth of the fungus in treated plates simultaneously.

Observations

....

....

....

....

Additional Reading

Hector, T and James A.M. (1986). Assessement and control of Biochemical Methods, John Wiley and Sons. Chichester U.K.

Robert B (1982). The Merk manual of diagnosis and therapy. Merk sharp and Dohme Research Laboratories. Division of Merck and Co. Inc. Rahway, NJ. USA

Taylor, E. (1989). Clinical Chemistry, Wiley Interscience, New York U.S.A.

❑❑❑

Unit -11

Plant Hormones

Thiamann (1948) designated the plant hormones by the term 'phytohormones' in order to distinguish them from animal hormones. He defined a phytohormone as "an organic compound produced naturally in higher plants, controlling growth or other physiological functions at a site remote from its place of production and active in minute amounts."

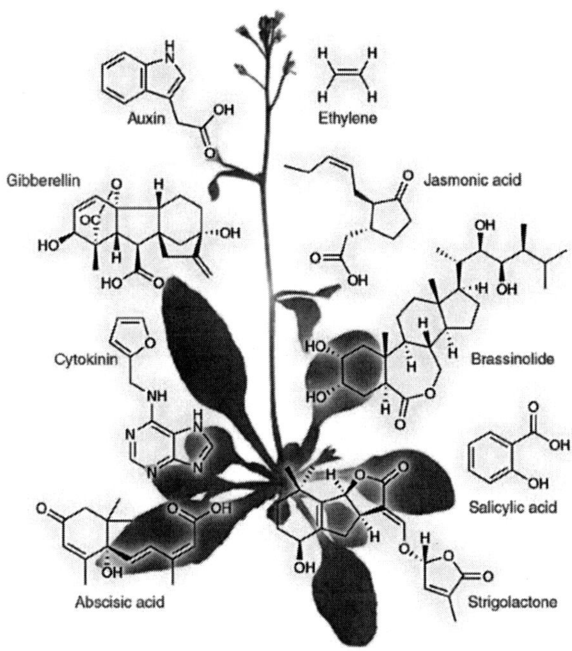

Fig. 11.1 : Structure of Different Plant Hormones

According to Johannes van Overbeek (1950), the plant hormones are defined as "organic compounds which regulate plant physiological process – regardless of whether these compounds are naturally occurring and/or synthetic; simulating and/or inhibitory; local activators or substances which act at a distance from the place where they are formed."

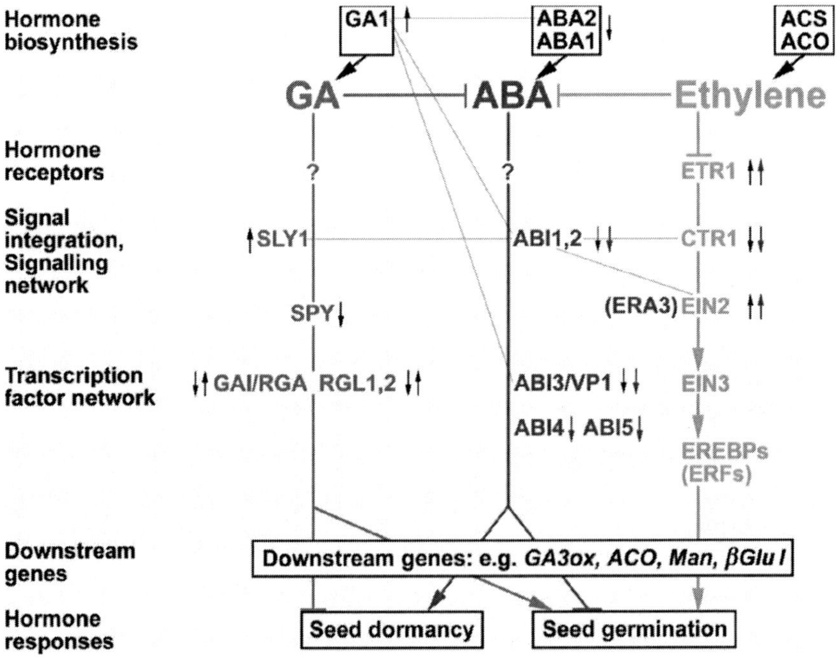

Fig. 11.2 : Hormonal interactions during seed dormancy release and germination (*Source:* Kucera, Cohn, Leubner-Metzger (2005) "Plant hormone interactions during seed dormancy release and germination". Seed Science Research 15: 281-307.)

- *"Great scientific discoveries have been made by mean seeking to verify quite arroneous theories about the nature of things."*

  Aldous Huxley

- *"Keep your heart open to dreams for a long as there's dream, there is hope, and as long as there is hope, there is joy in living."*

  Anonymous

Experiment – 1 : Identify the Indoleacetic Acid Present in the Given Sample

Principle

IAA is reached with trifluoracetic acid and acetic anhydride to convert it into indole-α-pyrone which is measured fluorimetrically.

Requirements

Reagents: Methanol (re-distilled), Trifluoroacetic acid – Acetic anhydride reagent: Mix equal volume of each liquid, precooled to 0°C. Freshly prepare and store on ice until use, methanol, phosphoric acid, petroleum ether, diethyl ether, K_2HPO_4.

Glass/Plastic wares: Beakers, test tubes, measuring cylinder, Buchner funnel.

Equipments: Fluorimeter: Primary filter 440 nm and secondary filter 490 nm (for low concentration measurement). Micropipettes, pestle mortar, refrigerator.

Extraction of IAA

Freeze a known quantity (5 g) of plant material in liquid nitrogen and grind to a fine powder using a pestle and mortar. Continue grinding with 10 ml methanol (redistilled) to a fine suspension. Filter the homogenate through a G4 glass under suction into a 100 ml flask. Extract the material on the filter twice by adding 10 ml methanol and then once with 5 ml. evaporate the filtrate in a rotary evaporator at 30°C to an aqueous residue.

To the aqueous residue add 10 ml of cold 0.5 M K_2HPO_4 solution so that pH reaches to about 8.5. Transfer to a suitable separating funnel and shake with 10 ml light petroleum ether. Repeat this step again and then shake with 10 ml light petroleum ether. Repeat this step again and then shake with 10 ml diethyl ether each time discards the lipid fraction. Adjust the aqueous layer to pH 3 by adding about 3 ml of 2.8 M phosphoric acid. Extract IAA with 10 ml diethyl ether.

The 10 ml diethyl ether is then extracted with 10 ml cold 50 mM K_2HPO_4 solution. The pH of the solution is adjusted to 3 with phosphoric acid (0.28 M) and the IAA is passed into a final 10 ml diethyl ether. The ether is then evaporated in a few min under reduced pressure. Dissolve the residue in a known volume (5 ml) of cold redistilled methanol.

Procedure

- Pipette out 1 ml of the above methanolic extract each in four different test tubes.
- To each tube add 1 ml of methanol containing 0, 10, 20 or 30 ng of IAA respectively.
- Dry the contents in each tube completely under reduced pressure and cool to 0°C.
- To each flask add 0.2 ml of ice-cold trifluroacetic acid – acetic anhydride reagent and mix.
- Place the tubes on ice for exactly 15 min to ensure the complete conservation of IAA into indole-α-pyrone. Stop the reaction by adding 3 ml water.
- A blank may be prepared occasionally by adding first 3 ml water to one of four aliquots and 0.2 ml reagent after 15 min.
- Take the readings in a fluorimeter using an extraction at 440 nm and emission at 490 nm for low concentration samples.
- Calculate the amount in unknown.

Notes :

1. *The reagent components and the reagent should be cooled on ice before use. Reaction should also be carried out on ice to avoid irregular results.*
2. *Chlorophylls and carotenoids interfere with the measurement. Avoid these contaminations during extraction be collecting only clear layers.*

Observations

....

....

....

....

Experiment – 2 : Estimation of Ethylene in the Plant Material

Principle

The ethylene is measured in a gas chromatograph based on the adsorption principle on activated silica gel or poropak.

Requirements

- Conical flasks or Cylinders with facility to seal with Rubber Gaskets
- Air-tight Syringes
- Ethylene Gas
- GLC

Column	-	Pretested 10% silar 10°C on gas chrom Q 100 to 200 mesh (6′ x ¼″ – internal diameter 4 mm).
Detector	-	FID
Injector Temp.	-	280°C
Detector Temp.	-	250°C
Carrier gas	-	Nitrogen at 50 ml/min
Column/oven temp.	-	165°C

Procedure

- Place the fruits in conical flasks or cylinders.
- Seal the mouth with rubber septum or gasket.
- Incubate for 1 h at 20°C.
- Withdraw gas samples with hypodermic syringe and inject into GLC.
- For standard, inject pure ethylene into empty conical flasks or cylinders of same volume and satisfy identical assay conditions. Remove the same volume of internal atmosphere as that of sample from the flask, inject into GLC and measure ethylene peak height.

Calculation

The quantity of ethylene produced is expressed as μl ethylene per hour per kg material.

Notes :

1. *Depending upon the size and quantum of fruit sample, select suitable conical flasks or cylinders.*
2. *For studying rate of ripening select fruits without any macroscopic wounds and remove surface dirt.*

Observations

....

....

....

....

Additional Reading

Abeles, F.B. (1973). Ethylene in plant biology. Academic Press, Inc., New York.

Evans, M.L. (1984). The action of auxin on plant cell elongation. CRC Crit. Rev. Plant Sci. 2: 317-365.

Lange, T(1998). The molecular biology of gibberellin synthesis. Planta. 204 : 409-419.

McGaw, B.A. Burch L.R. (1995). cytokinin biosynthesis and metobolism. In plant Hormones : Physiology, Biochemistry and Molecular biology. P.J. Davies ed. Kluwer Academic Publishers. Dordrecht pp. 98-117.

MoK, D.W.S. Mock M.C. eds (1994). Cytokinins Chemistry, Action and Function. CRC Press Boca Raton. FL.

Phillips, I.D.J. (1971). Introduction to the Biochemistry and Physiology of Plant Growth Hormones. McGraw Hill Book Co., New York.

Salisbury, F.B. and Rose, C.W. (1992). Plant Physiology. 4th edition. Wadsworth, Belmont, CA.

Went, F.W. and Thimann, K.V. (1937). Phytohormones. The Macmillion Co., New York.

❑❑❑

Appendices

Appendix – I

Definitions

1. ***Atomic weight:*** Atomic weight of an element is the relative weight of the atom on the basis of oxygen as 16.

   For example, Atomic weight of sodium is 23.

2. ***Molecular weight:*** The sum of the atomic weights of all the atoms in a molecule is its molecular weight.

   For example, Molecular weight of H_2SO_4 is 98, since-

Hydrogen	2 x 1	=	2
Sulphur	1 x 32	=	32
Oxygen	4 x 16	=	64
			98

3. ***Equivalent weight:*** Equivalent weight of a substance is the number of grams of the substance required to react with, replace or furnish one mole of H_2O^+ or OH^-.

   The equivalent weight of an acid is the weight that contains one atomic weight of acidic hydrogen i.e., the hydrogen that reacts during neutralization of acid with base.

For example, the equivalent weight of H_2SO_4 is 49. Since H_2SO_4 contains two replaceable hydrogens, equivalent weight is molecular weight/2 i.e., 98/2=49.

4. ***Percent Solution (w/v):*** One percent solution of a substance contains one gram of the substance in 100 mL of the solvent. If v/v is given, it means 1mL in 100 mL of solvent.

5. ***Molar solution (M):*** One molar solution of a substance contains one mole or one gram molecular weight of the substance in one litre of solution.

   For example, 1 M NaOH contains 40g sodium hydroxide in one litre solution. Likewise, one millimolar solution of a substance contains

6. ***Normal solution (N):*** One normal solution of a substance contains on equivalent or one gram equivalent weight of the substance in one litre solution (i.e., molecular weight divided by the hydrogen equivalent of the substance).

   For example, 1 N H_2SO_4 contains 49 g H_2SO_4 in one litre solution.

7. ***Buffer:*** A solution containing both a weak acid and its conjugate weak base whose pH changes only slightly on the addition of acid or alkali.

8. ***pH:*** pH is a value taken to represent the acidity or alkalinity of an aqueous solution. It is defined as logarithm of the reciprocal of the hydrogen ion concentration of the solution

   $$\text{i.e., pH} = \log \frac{1}{[H^+]}$$

9. ***Dilute acids:*** While preparing dilute acids, add slowly to water preferable under cooling.

10. ***Alkali solutions:*** When preparing concentrated alkali solutions (e.g., 40% NaOH) dissolve the alkali in distilled water under cooling.

11. ***Water:*** In biochemical experiments, it is advised to use distilled water for all purposes. So, water means distilled water in the methodologies given.

Concentration of Solutions

Molarity (M)	:	The number of moles of solute per litre of solution (The term molar is equivalent to mol l^{-1})
Molality (m)	:	The number of moles of solute per kg of solvent
% (w/v)	:	The weight in grams of a solute per 100 mL of solution
% (w/w)	:	The weight in grams of a solute per 100g of solution

Units of Length

1 micron = 1μ = 1 μm = 1 x 10^{-6} m = 1 x 10^{3} nm = 1 x 10^{4} Å

1 Å = 0.1 nm = 1 x 10^{-4} μm = 1 x 10^{-10} m

1nm = 10 Å = 1 x 10^{-3} μm = 1 x 10^{-9} m

Conversion Units of Temperature

Temp. in °F = (Temp. in °C x 1.8) + 32

Temp. in °C = (Temp. in °F – 32) x 5/9

Temp. in K = (Temp. in °F + 459.67) x 5/9

Units of Radioactivity

Curie (Ci)	=	That quantity of a radioactive substances in which 3.7 x 10^{10} atoms disintegrate per second. Thus 1Ci = 3.7 x 10^{10} dps = 2.22 x 10^{12} dpm.
Becquerel (Bq)	=	That quantity of a radioactive substance in which one atom disintegrate per second. Thus 1 Ci = 3.7 x 10^{10} Bq 1 Bq = 1 dps = 60 dpm

Some Isotopes and their Properties

Isotope	Type of decay	Half-life	Average energy
^{3}H	β^-	12.3 *y*	0.0055
^{14}C	β^-	5570 *y*	0.05
^{32}P	β^-	14.5 *d*	0.70
^{35}P	β^-	87.4 *d*	0.049
^{36}Cl	β^-	3.07×10^5 *y*	0.30

y = year; d = day

Appendix – II

Preparation of Buffers

1. Acetate Buffer

Stock Solutions

A : 0.2 M solution of acetic acid (11.55 mL in 1000 mL)

B : 0.2 M solution of sodium acetate (16.4 g of C_2H_2Na or 27.2g of $C_2H_3O_2Na.3H_2O$ in 1000 mL).

x mL of A, *y* ml of B, diluted to a total of 100 mL

x	y	pH
46.3	3.7	3.6
44.0	6.0	3.8
41.0	9.0	4.0
36.8	13.2	4.2
30.5	19.5	4.4
25.5	24.5	4.6
20.0	30.0	4.8
14.8	35.2	5.0
10.5	39.5	5.2
8.8	41.2	5.4
4.8	45.2	5.6

2. Boric Acid-Borax Buffer

Stock Solutions

A : 0.2 M solution of boric acid (12.4g in 1000mL)

B : 0.05 M solution of borax (19.5g in 1000mL; 0.2 M in terms of sodium borate)

50 mL of A, *x* mL of B, diluted to a total of 200 mL

x	pH
2.0	7.6
3.1	7.8
4.9	8.0
7.3	8.2
11.5	8.4
17.5	8.6
22.5	8.7
30.0	8.8
42.5	8.9
59.0	9.0
83.0	9.1
115.0	9.2

3. Carbonate-Bicarbonate Buffer

Stock Solutions

A : 0.2 M solution of anhydrous sodium carbonate (21.2 g in 1000 ml)

B : 0.2 M solution of sodium bicarbonate (16.8 g in 1000 mL) *x* ml of A, *y* ml of B, diluted to a total of 200 ml

x	y	pH
4.0	46.0	9.0
7.5	42.5	9.3
9.5	40.5	9.4
13.0	37.0	9.5
16.0	34.0	9.6
19.5	30.5	9.7
22.0	28.0	9.8
25.0	25.0	9.9
27.5	22.5	10.0
30.0	20.0	10.1
33.0	17.0	10.2
35.5	14.5	10.3
38.5	11.5	10.4
40.5	9.5	10.5
42.5	7.5	10.6
45.0	5.0	10.7

4. Citrate Buffer

Stock Solutions

A : 0.1 M solution of citric acid (21.01 g in 1000 ml)

B : 0.1 M solution of sodium citrate (29.41 g $C_6H\text{-}_5O_7Na_3.2H_2O$ in 1000 ml) *x* ml of A, *y* ml of B, diluted to a total of 100 ml.

x	y	pH	x	y	pH	x	y	pH
46.5	3.5	3.0	33.0	17.0	4.0	18.0	32.0	5.2
43.7	6.3	3.2	31.5	18.5	4.2	16.0	34.0	5.4
40.0	10.0	3.4	28.0	22.0	4.4	13.7	36.3	5.6
37.0	13.0	3.6	25.5	24.5	4.6	11.8	38.2	5.8
35.0	15.0	3.8	23.0	27.0	4.8	9.5	40.5	6.0
			20.5	29.5	5.0	7.2	42.8	6.2

5. Glycine-HCl Buffer

Stock Solutions

A : 0.2 M glycine (15.01 g in 1000 ml)

B : 0.2 N HCl

25 ml of A, x ml of B diluted to a total of 100 ml

x	pH
22.0	2.2
16.2	2.4
12.1	2.6
8.4	2.8
5.7	3.0
4.1	3.2
3.2	3.4
2.5	3.6

6. Phosphate Buffer

Stock Solutions

A : 0.2 M solution of monobasic phosphate (27.8 g in 1000 ml)

B : 0.2 M solution of dibasic sodium phosphate (53.65 g of $Na_2HPO_4.7H_2O$ or 71.7 g of $Na_2HPO_4.12H_2O$ in 1000 ml)

x ml of A, y ml of B, diluted to a total of 200 ml

x	y	pH	x	y	pH
93.5	6.5	5.7	45.0	55.0	6.9
92.0	8.0	5.8	39.0	61.0	7.0
90.0	10.0	5.9	33.0	67.0	7.1
87.7	12.3	6.0	28.0	72.0	7.2
85.0	15.0	6.1	23.0	77.0	7.3
81.5	18.5	6.2	19.0	81.0	7.4
77.5	22.5	6.3	16.0	84.0	7.5
73.5	26.5	6.4	13.0	87.0	7.6
68.5	31.5	6.5	10.5	89.5	7.7
62.5	37.5	6.6	8.5	91.5	7.8
56.5	43.5	6.7	7.0	93.0	7.9
51.0	49.0	6.8	5.3	94.7	8.0

7. Tris (Hydroxymethyl) Aminomethane (Tris. HCl) Buffer

Stock Solutions

A : 0.2 M solution of Tris (hydroxymethyl) aminomethane (24.2 g in 1000 ml)

B : 0.2 N HCl

50 ml of A, *x* ml of B, diluted to a total of 200 ml

x	pH
5.0	9.0
8.1	8.8
12.2	8.6
16.5	8.4
21.9	8.2
26.8	8.0
32.5	7.8
38.4	7.6
41.4	7.4
44.2	7.2

8. Glycine-NaOH Buffer

Stock Solutions

A : 0.2 M solution of glycine (15.01 g in 1L)

B : 0.2 M NaOH

50 ml of A + x ml of B, diluted to a total of 200 ml

x	pH
4.0	8.6
6.0	8.8
8.8	9.0
12.0	9.2
16.8	9.4
22.4	9.6
27.2	9.8
32.0	10.0
38.6	10.4
45.6	10.6

Appendix – III

Preparation of Standard Solutions

1. Standard NaOH solution (0.1M/0.1N)

Reagents

1. Approx 0.1N NaOH solution: Prepare a saturated solution of sodium hydroxide as follows. Add in small portions about 100g of AR Grade NaOH to 100ml of water in a flask, stopper and allow to stand for a few days or until the Na_2CO_3 settles to the bottom. The clear solution which is carbonate-free contains about 50g of NaOH/100ml (approx 50% solution or 12.5M). From this solution, pipette out 5.3 ml of clear supernatant liquid and transfer to a 1.0 litre rubber stoppered bottle and dilute upto the mark.
2. Standard HCl solution (0.1N): Prepare and standardize as described before.
3. Phenolphthalein solution: Dissolve 0.5g of phenolphthalein in 100ml of 95% alcohol.

Method

Using a pipette, transfer 20ml of standard HCl solution into 150ml beaker. Add 25ml of water and one drop of phenolphthalein indicator. Titrate with the sodium hydroxide (approx 0.1N) until a faint pink colour is obtained. Repeat titration 2-3 times. Calculation the exact normality of NaOH using the relationship $N_1 V_1 = N_2 V_2$.

2. Standard sodium thiosulphate solution (0.1N)

Reagents

1. Approx 0.1N $Na_2S_2O_3.5H_2O$ solution: Dissolve 24.82g of sodium thiosulphate in freshly boiled and cooled water and dilute to 1.0 L.
2. 0.1N potassium dichromate: Dissolve 4.903g of potassium dichromate in water and dilute to 1.0L.
3. Starch indicator: To 1.0g of starch add 10ml of water and pour the suspension into 90ml of boiling water. boil for 1-2 min and cool.
4. 15% Kl solution: Dissolve 15g of KI in water and dilute to 100ml.
5. HCl solution.

Method

Place in glass-stopped flask 20ml of 0.1N potassium and 10ml of 15% KI solution. Add 5ml of HCl. Dilute with 100ml of freshly boiled and cooled water and titrate against the 0.1N $Na_2S_2O_3.5H_2O$ until the yellow color has almost disappeared. Add a few drops of starch indicator and continue to add NaS_2O_3 until the blue color just disappears. Calculate the exact normality of $Na_2S_{2-}O_3$ using the relationship $N_1 V_1 = N_2 V_2$.

3. Standard H_2SO_4 solution (0.05M/0.1N)

Reagents

1. Approx 0.05M/0.1N H_2SO_4 solution: Dilute 3.0ml of pure concentrated H_2SO_4 solution (sp. Gr. 1.84, 96%) to 1.0L with water in a volumetric flask and mix thoroughly.
2. Anhydrous Na_2CO_3 solution (0.1N): Prepare as described under standard HCl.
3. Methyl orange indicator: Prepare as described under standard HCl.

Method

Fill a clean, dry burette with H_2SO_4 solution (0.05M/0.1N) and titrate against a known volume (10ml) of anhydrous Na_2CO_3 solution using 2 drops of methyl orange as an indicator until an orange or faint pink color end point is obtained. Repeat titration 2-3 times.

Calculation

Calculation the exact normality of H_2SO_4 using the relationship, $N_1V_1 = N_2V_2$ as described before.

4. Standard HCl Solution (0.1 M or 0.1 N)

Reagents

1. Approx 0.1M/0.1N HCl solution: Dilute 9.0ml (sp. Gr. 1.18, 35%) or 8.5ml (sp. gr. 1.19, 37%) of pure concentrated HCl to 1.0 litre with distilled water in a volumetric flask. Invert several times and transfer to a clean, dry bottle.
2. Anhydrous Na_2CO_3 solution (0.1N): Dry the pure anhydrous Na_2CO_3 on a watch glass in the oven at 110^0C for an

quantitatively to a beaker and dissolve in about 20ml of water. Transfer to 100ml volumetric flask with washings, up to the mark and mix.

3. Methyl orange indicator solution: Dissolve 0.5g of methyl orange in 1.0L of water.

Method

Fill a clean, dry burette with HCl solution (approx 0.1M/0.1N) and titrate against a known volume (10ml) of sodium carbonate (0.1N) using 2 drops of methyl orange indicator until an orange or faint pink color end point is obtained. Note the titre value and repeat titration 2-3 times.

Calculation

Using $N_1V_1 = N_2V_2$ relationship, the exact normality of HCl can be calculated.

Where, N_1 - Normality of HCl to be calculated
N_2 - Normality of Na_2CO_3 = 0.1N
V_1 - Volume of HCl (titre value)
V_2 - Volume of Na_2CO_3 (10ml)

Note :

Methyl red may also be used in place of methyl orange as an indicator provided the CO_2 in solution is expelled by boiling before the end point is reached.

5. Standard Hydrogen Peroxide Solution (0.1N)

Reagents

1. 0.1N H_2O_2: Take about 7.0g of chemically pure H_2O_2 solution (ca. 7.0 P.C.) in a 500 ml measuring flask and dilute to the mark with water. Shake, keep in a dark, cool place.
2. Standard $Na_2S_2O_3$ solution (0.1N): Prepare and standardize as described before.
3. Standard potassium permanganate solution: Prepare and standardize as described before.

Method

1. Standardization with thiosulphate solution

Take 25ml of H_2O_2 in a 250ml flask, treat with 10ml of 4N H_2SO_4, 1.0g of KI and 4 drops of 3% ammonium molybdate to catalyze the reaction. Titrate against standard sodium thiosulphate using starch as an indicator until the blue color just disappears.

2. Standardization with potassium permanganate

Take 25ml of H_2O_2 in a 250ml flask, titrate with 0.1N permanganate solution in presence of 10ml of dilute H_2SO_4 (1:5) until a light permanent pink color is obtained.

6. Standard Potassium Permanganate Solution (0.1N)

Reagents:

1. Potassium permanganate solution (0.1N approximate) : Dissolve about 3.2g of pure crystals of potassium permanganate in 200ml of hot water and allow to stand for 60 min at about 100^0C. The mouth of the container is well closed with a glass-stopper to disallow any addition of dust particles into the permanganate solution. The solution is then filtered after 2 days standing through either a sintered filter or a filter of glass wool but never filter through a filter paper because the latter reduces the strength of permanganate considerably. The solution is kept in dark because of its sensitivity to light.
2. Oxalic acid ($H_2C_2O_4.2H\text{-}_2O$) or sodium oxalate $(COONa\text{-})_2$ (0.1N): Dissolve 6.3g of pure crystals of oxalic acid in 200ml of water and transfer to a litre flask. Make up with water or slowly heat 6.7g of pure sodium oxalate salt in a platinum crucible until the flame of carbon monoxide formed disappears. Dissolve the contents of crucible with water and make up to 1.0L.
3. 4N H_2SO_4.

Method

Dilute 10ml of standard oxalic acid or oxalate solution to 20ml water and treat with 15ml of 4N H_2SO_4. Heat the solution to 70-80^0C and titrate against permanganate solution until the solution becomes slightly pink. The color is permanent. Calculate the normality of permanganate solution using the relationship $N_1V_1 = N_2V_2$.

7. Standard Iodine Solution (0.1N)

Reagents

1. Iodine solution (0.1N): Dissolve 12.7g of iodine (A.R.) and 40g of iodate-free potassium iodide in 50ml of water. Stir the mixture until the whole of iodine dissolves completely. Transfer it to a

litre flask and make up to the mark with water. Keep in a dark and cool place.

2. Standard arsenious oxide solution (0.1N): Dry about 8.0g of arsenious oxide (As_2O_3) for 2h at 105-110^{0}C in a platinum crucible and cool in a desiccator. Wigh accurately 4.945g of dried sample and dissolve in 10mlof 4N NaOH solution. The solution may be heated if required. Transfer to a litre flask and treat with 4 drops of phenolphthalein. The alkalinity of medium is neutralized by carefully adding just enough dilute HCl (disappearance of pink color). Then mix the solution with another 50ml of dil. HCl and make up the volume to the litre mark.
3. Starch solution: Dissolve 1.0g of starch and 5g of $HgCl_2$ in 1.0ml of water and add the suspension to 500ml of boiling water. Boil until the clear solution is obtained and store in a glass-stoppered bottle.

Method

Take 25ml of standard arsenious oxide solution in 250ml of flask and mix with 50ml of water, 1.0g of sodium bicarbonate and 1.0ml of starch solution. Titrate against prepared I_2 solution until blue color appears. Calculate the normality of I_2 using the relationship $N_1V_1 = N_2V_2$.

Note :

Best results are obtained if the starch solution is added just before the end point is reached.

8. Standardization of the Fehling's Solution

Reagents

1. Fehling's solution A (copper sulphate solution): Dissolve 34.639g of $CuSO_4.5H_2O$ (pure recrystallized) in water, dilute to 500ml and filter.
2. Fehling's solution B (alkaline tartrate solution): Dissolve 173g of Rochelle salt (sodium potassium tartrate) and 50g of NaOH in water, dilute to 500ml. Allow to stand for 2 days and filter. Mix equal volumes (5ml each) of both immediately before use.
3. Standard glucose solution (0.5%): Dissolve accurately 0.5g of pure anhydrous glucose powder in water and dilute to 100ml with water.
4. 1.0% Methylene blue: Dissolve 1.0g in 100ml 95% alcohol.

Method

Into a conical flask, pipette out 5.0ml of each of Fehling's solutions A and B. Dilute with about 10ml of water and add 3 glass beads to it. Keep the flask over a wire gauze or hot plate and bring to boil. Add from the burelle the standard glucose solution slowly, 1ml at a time till the color just turns from blue to red. Then add 3-4 drops of 1.0% methylene blue indicator which gives a blue color to the solution and continue to add the glucose solution drop-wise till a brick red color end point of solution is obtained.

Again in another conical flask, take 5.0ml of each of the Fehling's solutions A and B, dilute with about 10ml of water and add 3 glass beads into it. Boil the contents as above and add from the burette the standard glucose solution 2.0ml less than the quantity required in the first titration. Then add 3-4 drops of methylene blue indicator and continue the titration till the brick red colored end point is obtained. Repeat the whole process of titration 3-4 times and calculate the mean of the last 3 readings (excluding the first reading). From the mean titration reading, calculate the amount of anhydrous glucose corresponding to 10ml of the mixed Fehling's solution A and B.

Note :

The titration should be completed within total boiling time of about 3 min. Note down the first burette reading.

Calculation

Let 'a' ml be the mean titration reading.

10ml of Fehling's A & B together = 'a' ml of std. glucose solution.

Now, 100ml of std. glucose solution = 0.5g of glucose

$\therefore$ 'a' ml of std. glucose solution = $0.5 \times \frac{a}{100}$ g of glucose

$\therefore$ 10ml of Fehling's solutions = $0.5 \times \frac{a}{100}$ g of glucose

A and B contain

9. Standard Sodium Nitrite Solution (0.1N)

Method

50ml of 0.1N permanganate solution and 5ml of dilute H_2SO_4 (1:5) are mixed together and treated with 25ml of approx 0.1N $NaNO_2$

solution. The mouth of flask is closed with a glass stopper and allowed to stand for 15 min. Then the excess of permanganate is titrated back by procedure described under preparation of standard potassium permanganate solution or iodometrically after adding 2.0g of KI and then titrating with standard thiosulfate solution.

Calculation

Volume of 0.1N $KMnO_4$ consumed by nitrite

= (50-volume of 0.1N standard solution used for back titration)

= actual volume of permanganate = V_1

Calculate normality of $NaNO_2$ using the relationship $N_1V_1 = N_2V_2$

Appendix – IV

Molecular Weights of Selected Laboratory Chemicals

Acrylamide	71.08	Phenol	94.11
Ammonium Acetate	77.08	Phenylmethylsulfonyl fluoride	174.2
Ammonium Bicarbonate	79.06	Polyethylene glycol	c. 8000
Ammonium Chloride	53.49	Polyvinylpyrrolidone	40,000
Ammonium Persulfate	228.2	Potassium Acetate	98.14
Ammonium Sulfate	132.14	Potassium Chloride	74.55
Ampicillin	371.4	Potassium Phosphate, anhydrous, monobasic	136.1
ATP, disodium salt	605.19	Potassium Phosphate, trihydrate, dibasic	228.2
dATP, disodium salt	535.1	Sodium Bicarbonate	84.01
Bis	154.17	Sodium Azide	65.01
Boric Acid	61.83	Sodium Bicarbonate	84.01
Calcium Chloride	147.0	Sodium chloride	58.45
Cesium Chloride	168.37	Sodium hydroxide	40.0
Chloramphenicol	323.1	Sodium iodide	149.89
Citric Acid	294.1	Sodium Dodecyl Sulfate	288.38
dCTP, sodium salt	467.2	Sodium Phosphate, monohydrate	137.99
Dithiothreitol	154.24	Sodium Phosphate, dibasic	142.0
EDTA, disodium salt	372.24	Spermidine, trihydrochloride	254.6
Ficoll type 400	400,000	Spermine, tetrachloride	348.2
Glucose	180.2	Sucrose	342.3
Glutamic acid	147.1	TEMED	116.2
dGTP, sodium salt	507.2	Thiamine	337.3
Glycine	75.07	Thymidine	242.2
Hydroxyquinoline	145.2	Tris	121.1
IPTG	238.31	Tris-HCl	157.6
2-mercaptoethanol	14.26 M	dTTP, sodium salt	482.2
Magnesium Chloride, hexahydrate	203.30	Urea	60.06
Magnesium Sulfate, anhydrous	120.39	X-gal	408.64

Appendix – V

The French **Système International d'Unités** (the SI system) is accepted convention for all units of measurement.

The SI unit of mass (weight) is given below:

Table 1.1: SI Units of Mass

Weight	Symbol	Multiple of gram	Lower equivalent
megagram	M	10^6	1000 kilogram
kilogram	K	10^3	1000 gram
gram	g	1	1000 milligram
milligram	m	10^{-3}	1000 microgram
microgram	μ	10^{-6}	1000 nanogram
nanogram	n	10	1000 picogram
picogram	p	10	1000 femtogram

Table 1.2 : SI units –basic and derived units

Quantity	SI unit	Symbol (basic SI units)	Definition of SI unit	Equivalent in SI units
<u>**Basic units**</u>				
Length	metre	m		
Mass	kilogram	kg		
Time	second	s		
Electric current	ampere	A		
Temperature	kelvin	K		
Luminous intensity	candela	cd		
Amount of substance	mole	mol		
<u>**Derived units**</u>				
Force	newton	N	$kgms^{-2}$	Jm^{-1}
Energy, work, heat	joule	J	kgm^2s^{-2}	Nm
Power, radiant flux	watt	W	kgm^2s^{-2}	Js^{-1}
Electric charge, quantity	coulomb	C	As	JV^{-1}
Electric potential difference	volt	V	$kgm^2s^{-3}A^{-1}$	JC^{-1}
Electric resistance	ohm	&!	$kgm^2s^{-2}A^{-2}$	VA^{-1}
Pressure	pascal	Pa	$Kgm^{-1}s^{-2}$	Nm^{-2}
Frequency	hertz	Hz	s^{-1}	

Contd.

Magnetic flux density	tesla	T	$kgs^{-2}A^{-1}$	Vsm^{-2}
Other units based on SI				
Area	square metre	m^2		
Volume	cubic metre	m^3		
Density	kilogram per cubic metre	Kgm^{-3}		
Concentration	Mole per cubic metre	mol m^{-3}		

Table 1.3 : SI units – conversion factors for non-SI units

Unit	Symbol	SI equivalent
Avogadro constant	L or N_A	6.022×10^{23} mol^{-1}
Faraday constant	F	9.648×10^4 $Cmol^{-1}$
Planck constant	h	6.626×10^{-34} Js
Universal or molar gas constant	R	8.314 $JK^{-1}mol^{-1}$
Molar volume of an ideal gas at s.t.p.		22.41 dm^3mol^{-1}
Velocity of light in a vacuum	c	2.997×10^8 ms^{-1}
Energy		
calorie	cal	4.184 J
erg	erg	10^{-7} J
electron volt	eV	1.602×10^{-19} J
Pressure		
atmosphere	atm	101325 Pa
bar	bar	10^5 Pa
Millimeters of Hg	mm Hg	133.322 Pa
Temperature		
centigrade	^{0}C	(t ^{0}C + 273.15) K
Fahrenheit	^{0}F	(t ^{0}F – 32)5/9 + 273.15 K
Length		
Ångström	Å	10^{-10} m
inch	in	0.0254 m
Mass		
pound	lb	0.4536 kg

Note: s.t.p., standard temperature and pressure.

Table 1.4 : Common unit prefixes associated with quantitative terms

Multiple	Prefix	Symbol	Multiple	Prefix	Symbol
10^{24}	yotta	Y	10^{-1}	deci	d
10^{21}	zetta	Z	10^{-2}	centi	c
10^{18}	exa	E	10^{-3}	milli	m
10^{15}	peta	P	10^{-6}	micro	µ
10^{12}	tera	T	10^{-9}	nano	n
10^{9}	giga	G	10^{-12}	pico	p
10^{6}	mega	M	10^{-15}	femto	f
10^{3}	kilo	k	10^{-18}	atto	a
10^{2}	hecto	h	10^{-21}	zepto	z
10^{1}	deca	da	10^{24}	yocto	y

Table 1.5 : Interconversion of non – SI and SI units of volume

Non-SI unit	Non-SI subunit	SI subunit	SI unit
1 litre (l)	10^3 ml	= 1 dm^3	= 10^{-3} m^3
1 mililitre (ml)	1 ml	= 1 cm^3	= 10^{-6} m^3
1 microlitre (µl)	10^{-3} ml	= 1 mm^3	= 10^{-9} m^3
1 nanolitre (nl)	10^{-6} ml	= 1 nm^3	= 10^{-12} m^3

Table 1.6 : Interconversion of mol, mmol and µmol in different volumes to give different concentrations

Molar (M)	Millimolar (mM)	Micromolar (µM)
1 mol dm^{-3}	1 mmol dm^{-3}	1 µmol dm^{-3}
1 mmol cm^{-3}	1 µmol cm^{-3}	1 nmol cm^{-3}
1 µmol mm^{-3}	1 nmol mm^{-3}	1 pmol mm^{-3}

Appendix – VI

Constants of Acids and Bases

Acid or Base	Formula	Molecular wt.	Commercial concentrated reagent		
			Specific gravity	% by weight	Molarity (M)
Acetic acid	CH_3COOH	60.1	1.05	99.5	17.4
Ammonium Hydroxide	NH_4OH	35.0	0.89	28	14.8
Formic acid	$HCOOH$	46.0	1.20	90	23.4
Hydrochloric acid	HCl	36.5	1.18	36	11.6
Nitric acid	HNO_3	63.0	1.42	71	16.0
Perchloric acid	$HClO_4$	100.5	1.67	70	11.6
Phosphoric acid	H_3PO_4	80.0	1.70	85	18.1
Sulphuric acid	H_2SO_4	98.1	1.84	96	18.0

Appendix – VII

Densities of liquids

If the density of the solution is known, it is possible to calculate the water volume (weight) that is necessary for its preparation. It is very convenient, because otherwise phrase "add water to the final volume..." forces you to use the graduated cylinder for the preparation of solution (calibration of all other laboratory glassware is only approximate).

A list of densities of some chemical substances is given below

Substance	Density [g/ml]	Substance	Density [g/ml]
Acetic acid glacial	1.049	Mercaptoethanol (ß)	1.114
Acetic acid 36%	1.049	$MgCl_2$, 1M	1.073
Acetone	0.788	$MgSO_4$, 1M	1.112
AcOK, 5.0M 1.0M	1.2181.046	$MnCl_2$, 1M	1.099
AcONa, 3.0M 2.5M	1.1211.147	Nitric acid, $HNO_3$71%	1.42
$AcONH_4$, 10M	1.073	NaCl, 5M	1.186
Acrylamid:BIS, 40%	1.044	NaOH, 50%	1.53
Ammonium hydroxide, NH_4OH, 28%	0.89	NaOH, 10M	1.329
Butanol (n)	0.81	$NaxHPO_4$, 1M,pH~7	1.104
$CaCl_2$, 1M	1.086	Nonident P-40	1.064
Chloroform	1.48	Octane	0.71
Denhardt sol., 50X	1.004	PEG 6000, 20%40%	1.0411.075
DMSO	1.100	PEG/TAE, (15%/1X)	1.032
EDTA, 0.5M	1.096	Phosphoric acid H_3PO_4, 85%	1.70
Ethanol, anhydr 96% 70%	0.790.800.868	Phenol	1.054
Formaldehyde, 37%	1.09	Sulfuric acid H_2SO_4, 96%	1.84
Formamid	1.133	Sarcosyl, 10% (w/v)30% (w/v)	1.0171.027
Glucose, 10%(w/v)2M	1.0401.13	SDS, 10%	1.018
Glycerol	1.25	SSC, 20X30X	1.1631.245
HCl, 37% (12.07M)	1.19	SSPE, 20X	1.136
Isoamyl alcohol	0.813	TAE, 50X	1.088
Isobutanol	0.871	TBE, 10X	1.070

Contd.

Isopropanol	0.785	Tris.Cl, 1M, pH 7-82M, pH 7-9	1.0411.10
KCl, 0.25M	1.044	TritonX-100, 10%	1.016
KOH, 5.0M	1.217	TritonX-100, 100%	1.06
LB	1.024	Tween 20, 100%	1.108 (1.101)
LiCl, 9M8M	1.1941.172	$ZnCl_2$, 1M	1.111

Appendix – VIII

Weights and Measures

Decimal fractions

Fraction	Symbol	Prefix	Abbreviation
Thousandth	10^{-3}	Milli	M
Millionth	10^{-6}	Micro	μ
Billionth	10^{-6}	Millimicro (or nano)	mμ
Trillionth	10^{-12}	Micromicro (or pico)	μμ

Molar concentration

Mole fraction	Name	Abbreviation	Amount/L	Amount/ml
	Molar	M	1 mole/L	1 mmole/ml
10^{-3}M	Millimolar	mM	1 millimole/L	1 μmole/ml
10^{-6}M	Micromolar	μM	1 μmole/L	1 mμmole/ml
10^{-9}M	Nanomolar	nM	1 mμmole/L	1 μμmole/ml

Weights

Name	Abbreviation	Fraction of gram (g)
Gram	g	
Milligram	mg	10^{-3}
Microgram (gamma)	μg	10^{-6}
Nanogram	ng	10^{-9}
Picogram	pg	10^{-12}

Volume

Name	Abbreviation	Fraction of litre
Liter	L	
Milliliter	ml	10^{-3}
Microliter	μl	10^{-6}
(lambda)	(λ)	

Length

Name	Abbreviation	Fraction of meter
Centimeter	cm	10^{-2}
Millimeter	mm	10^{-3}
Micron	μ	10^{-6}
Millimicron	mμ	10^{-9}
Angstrom unit	Å	10^{-10}

Temperature

Temp. in °F = (Temp. in °C x 1.8) + 32

Temp. in °C = (Temp. in °F - 32) x 5/9

Temp. in K = (Temp. in °F + 459.67) x 5/9

Appendix – IX

pK values of Acids/Bases

Acids/Bases	pK values
Acetic acid	4.76
Ammonia	4.74
Barbitone	7.98
Barbituric acid	4.04
Boric acid	9.23
Carbonic acid	6.37 [pK_1]
	10.25 [pK_2]
Phosphoric acid	2.12 [pK_1]
	7.21 [pK_2]
	12.32 [pK_3]
Pyridine	8.64
Sulphuric acid	0.4 [pK_1]
	1.92 [pK_2]

Appendix – X

Preparation of Indicators

1) 0.04% Bromophenol blue: Grind 0.1g of dry bromophenol blue powder in a mortar with 14.9 ml of 0.01M NaOH. Dilute to 250 ml with water.
2) 0.04% Bromothymol blue: Grind 0.1 g of bromothymol blue dry powder in a mortar with 16 ml NaOH. Dilute to 250 ml with water.
3) 0.5% Methyl red: Dissolve 0.5g of methyl red in 100 ml of 95% ethyl alcohol.
4) Methylene blue-methyl red mixed indicator: This solution consists of 1 part of 0.2% methyl red in alcohol and 1 part of 0.1% methylene blue in alcohol. The pH is at 5.4. The color is green in alkaline (pH 5.6), colorless at pH 5.4 (end point) and reddish pink in acid solution (pH 5.2).
5) 0.02% Methylene blue: Dissolve 0.2g of 'certified' methylene blue in 1L of water.
6) 0.04% Phenol red: Grind 0.1g of dry phenol red powder in a mortar with 28.2ml of 0.01M NaOH. Dilute to 250ml with water.
7) 0.5% Phenolphthalein solution: Dissolve 0.5g phenolphthalein in 100ml of 95% alcohol.
8) 1% Starch: Mix 1g of soluble starch with a little water, stir and add to 100ml of boiling water.
9) 0.1% Bromocresol green: Dissolve 20mg in 20ml water.
10) 0.5% Congo red: Dissolve 5g of congo red in 900ml of water and dilute to a litre with 95% alcohol.
11) Methyl orange: Dissolve 0.5g of methyl orange in 1.0L of water.
12) Litmus: Dissolve 1.0g of azolitmin in 1.0L of water.
13) Potassium chromate: Dissolve 5.0g of potassium chromate in 100ml of water.

Appendix – XI

pH ranges of Indicators

Indicator	pH Range	Color change
Acid cresol red	0.2-1.8	Red-yellow
Acid metacresol purple	1.2-2.8	Red-yellow
Thymol blue	1.2-2.8	Red-yellow
Benzo yellow	2.4-4.0	Red-yellow
Bromocresol green	3.8-5.4	Yellow-blue
Bromophenol blue	3.0-4.6	Yellow-blue
Methyl red	4.4-6.0	Red-yellow
Chlorophenol red	5.2-6.8	Yellow-red
Bromocresol purple	5.2-6.8	Yellow-purple
Bromothymol blue	6.0-7.6	Yellow-blue
Phenol red	6.8-8.4	Yellow-red
Cresol red	7.2-8.8	Yellow-red
Meta cresol purple	7.6-9.2	Yellow-purple
Thymol blue	8.0-9.6	Yellow-blue
Phthalein red	8.6-10.2	Yellow-red
Tolyl red	10.0-11.6	Red-yellow
Parazo orange	11.0-12.6	Yellow-orange
Acyl blue	12.0-13.6	Red-blue

Appendix – XII

Distribution of Pigments and their Absorption Peaks

Type of pigments cells in nm	Distribution in plant kingdom	Absorption peaks in	
Chl a	All green plants	435	670-680
Chl b	All green plants except diatoms, brown, red and blue green algae	480	650
Chl c	Diatoms and brown algae	-	645
Chl d	Some red algae	-	740
Protochlorophyll	Etiolated plants	-	-
Bacterioviridin	Green sulphur bacteria	-	750 (or 760)
Bacteriochlorophyll	Purple sulphur bacteria	800, 850 and 890	

The Carotenoids

Type of carotenoids	Distribution in Plant Kingdom	Characteristic absorption peaks, nm
	I. Carotenes	
α- Carotene	Many leaves, and certain algae. In red algae and a group of green algae called Siphonales	In hexane, at 420, 440, 470
β- Carotene	Main carotenes of all other plants	In hexane, at 425, 450, 480
γ- Carotene	Major carotenes of green sulphur bacteria; traces in some plants	In hexane, at 440, 460, 495
	II. Carotenols (Also called 'Xanthophylls')	
Luteol	Major carotenol of green leaves, green algae and red algae	In ethanol, at 425, 445, 475
Violaxanthol	Second major carotenol of leaves	In ethanol, at 425, 450, 475
Fucoxanthol	Major carotenol of diatoms and brown algae	In hexane, at 425, 450, 475
spirilloxanthol	Common in purple bacteria	In hexane, at 464, 490, 524

The distribution of Phycobillins in the plant kingdom

Type of phycobilins	Distribution in plant kingdom	Absorption peaks in nm
Phycocyanins	Mostly in blue green algae	In water and in vivo, at 618
Phycoerythrins	Mostly in red algae	In water and in vivo, at 490, 546, and 576
Allophycocyanins	Blue green and red algae	In phosphate buffer at pH 6.5, at 654

❑❑❑

Index

❑❑❑